Education in the Asia-Pacific Region: Issues, Concerns and Prospects

Volume 78

The purpose of this Series is to meet the needs of those interested in an in-depth analysis of current developments in education and schooling in the vast and diverse Asia-Pacific Region. The Series will be invaluable for educational researchers, policy makers and practitioners, who want to better understand the major issues, concerns and prospects regarding educational developments in the Asia-Pacific region.

The Series complements the Handbook of Educational Research in the Asia-Pacific Region, with the elaboration of specific topics, themes and case studies in greater breadth and depth than is possible in the Handbook.

Topics to be covered in the Series include: secondary education reform; reorientation of primary education to achieve education for all; re-engineering education for change; the arts in education; evaluation and assessment; the moral curriculum and values education; technical and vocational education for the world of work; teachers and teaching in society; organisation and management of education; education in rural and remote areas; and, education of the disadvantaged.

Although specifically focusing on major educational innovations for development in the Asia-Pacific region, the Series is directed at an international audience.

The Series Education in the Asia-Pacific Region: Issues, Concerns and Prospects, and the Handbook of Educational Research in the Asia-Pacific Region, are both publications of the Asia-Pacific Educational Research Association.

Those interested in obtaining more information about the Monograph Series, or who wish to explore the possibility of contributing a manuscript, should (in the first instance) contact the publishers.

Please contact Lay Peng, Ang (e-mail: laypeng.ang@springer.com) for submitting book proposals for this series.

Rupert Maclean, University of Tasmania, Hobart, Australia
Philip Wing Keung Chan, Monash University, Australia

Aysuhan Tuba Saral • Margarita Pavlova

Education for Refugees and Sustainable Green Urban Development

Building Skills through Non-formal Learning

Springer

Aysuhan Tuba Saral
Department of International Education
The Education University of Hong Kong
New Territories, Hong Kong SAR, China

Margarita Pavlova
Department of International Education
The Education University of Hong Kong
New Territories, Hong Kong SAR, China

ISSN 1573-5397 ISSN 2214-9791 (electronic)
Education in the Asia-Pacific Region: Issues, Concerns and Prospects
ISBN 978-981-95-1679-7 ISBN 978-981-95-1680-3 (eBook)
https://doi.org/10.1007/978-981-95-1680-3

This Springer imprint is published by the registered company Springer Nature Singapore Pte Ltd.
The registered company address is: 152 Beach Road, #21-01/04 Gateway East, Singapore 189721, Singapore

Acknowledgment

This work is conducted as a doctoral study at the Education University of Hong Kong by Aysuhan Tuba SARAL (Ph.D. student) and supervised by Dr Margarita PAVLOVA. Hence, we would like to thank the University Grants Committee (UGC) of Hong Kong for funding the study and the fieldwork.

The study presented in the book used resources developed by the project, namely *Creating an Impact: Setting up a Green Skills Hub at the EdUHK #03A68*, led by Dr Margarita Pavlova, and funded by the Education University of Hong Kong Central Reserve Allocation Committee (CRAC), between July 2017 and October 2019.

Special thanks to Branches of Hope (BoH), a local charity that serves Hong Kong refugees, asylum seekers, and human trafficking survivors, as well as their representative, Mr. Ng Tsz Nok, for their collaboration and facilitation of this study.

Thanks should also go to Mr. Innocent Mutanga, the founder of a social enterprise in Hong Kong that strives to re-brand blackness. We appreciate his earnest sharing of his journey, which he began as a refugee and became a social enterprise founder in the city. His resilience and achievements are inspiring to all of us.

We would like to thank Prof. Rupert Maclean, who was an external examiner during the thesis defence and recommended publishing this study as a book. We are grateful for your support.

Finally, we thank our participants. We are proud of them, and we hope this book is a testimony to their achievements and potential when given a chance.

Some parts of this book related to pedagogy and workshops are published as a book chapter elsewhere. Please see below:

Saral, A. T., & Pavlova, M. (2025). “17: Green skills for refugees in Hong Kong”. In *A Modern Guide to Refugee Education*. Cheltenham, UK: Edward Elgar Publishing. https://doi.org/10.4337/9781803928401.00026

Series Editors Introduction

This timely and original volume, *Education for Refugees and Sustainable Green Urban Development: Building Skills through Non-formal Learning*, authored by Aysuhan Tuba Saral and Margarita Pavlova, is a compelling addition to the Springer Book Series Education in the Asia-Pacific Region: Issues, Concerns and Prospects. Now marking over two decades of publication, this 78th volume deepens the series' longstanding commitment to equity, innovation, and the transformative power of education in an evolving regional and global context.

In an era defined by ecological precarity, rising urban populations, and unprecedented displacement, the authors direct our attention to a powerful yet underexplored intersection: the role of refugees and asylum seekers (RAS) in advancing sustainable urban development (SUD). By focusing on Hong Kong as a microcosm of these global dynamics, the book offers a rare and much-needed perspective on how environmental resilience and social inclusion can, and must, be pursued in tandem.

This book challenges prevailing policy and academic discourses that view refugees predominantly through lenses of vulnerability or cost. Instead, Saral and Pavlova assertively reposition RAS as *potential agents of sustainability.* The book offers a grounded and hopeful narrative of empowerment through education, framed by the concept of **generic green skills (GGS)**, a set of competencies critical for transitioning to low-carbon, environmentally sustainable societies.

What distinguishes this study is not only its conceptual innovation but also its methodological depth. Employing a robust qualitative design and an intervention-based research framework, the authors document a targeted workshop initiative for refugee women and their children, underpinned by the principles of participatory learning and social entrepreneurship. The multi-layered evaluation draws on Kirkpatrick's framework, resulting in actionable insights at personal, community, and policy levels.

Equally commendable is the book's commitment to centring refugee voices. Rather than assuming a top-down lens, the authors illuminate RAS experiences, values, and perspectives on sustainability. Through interviews, focus groups, and reflective practice, the participants' narratives emerge as essential data, challenging assumptions, informing pedagogy, and calling into question entrenched policy and social discourses.

The Hong Kong context, historically shaped by refugee flows but institutionally reluctant to recognize asylum seekers, provides a poignant backdrop. Saral and Pavlova deftly navigate the tensions between government sustainability initiatives and the exclusionary realities faced by RAS populations. Their analysis extends beyond critique, however, to offer a constructive roadmap for inclusive educational and policy reform.

Among the most significant contributions of this volume is its articulation of how green skills development can serve as a pathway to **well-being, dignity**, and **economic participation** for displaced populations. This reframing aligns with and extends Sustainable Development Goals, particularly those related to inclusive and quality lifelong education (SDG 4), decent work and economic growth (SDG 8), reduced inequalities (SDG 10), and sustainable cities and communities (SDG 11).

Education for Refugees and Sustainable Green Urban Development exemplifies the series' mission: to showcase contextually rich, theoretically informed, and practically impactful scholarship. It addresses urgent questions with rigour and clarity and makes a compelling case for why non-formal education must be seen as a legitimate and essential domain of policy innovation, especially in the context of forced migration and environmental crisis.

For scholars, this volume presents a replicable research model and a novel pedagogical framework. For practitioners and policymakers, it delivers a persuasive call to action, one that places marginalized voices at the heart of the green transition. For all concerned with the future of education in the Asia-Pacific and beyond, this book is a timely and important contribution.

We are honoured to include this volume in our series and invite readers and prospective contributors alike to join us in advancing a more inclusive, sustainable, and just vision of education.

This Springer Book Series offers an in-depth exploration of critical themes such as the integration of technology in education, reforms across primary, secondary, and teacher education, innovative assessment techniques, alternative educational models, and strategies to achieve universal quality education. It also highlights initiatives to promote active aging through lifelong learning. The series features case studies from diverse educational systems across the Asia-Pacific region, comparative cross-cultural and cross-national studies, and analyses of the sociology of teachers as a professional group, among other related topics. For more details, please visit the Springer Book Series website at http://www.springer.com/series/5888.

Each volume is crafted to engage a broad audience, including researchers, policymakers, educators, higher education students, and classroom teachers at all levels, as well as general readers with a keen interest in current educational trends and developments throughout the Asia-Pacific region.

The *Education in the Asia-Pacific Region: Issues, Concerns and Prospects* series is dedicated to examining the dynamic and complex educational landscape of this diverse region. Influenced by unique regional challenges and rapid societal changes, the series investigates how education interacts with workforce development, employment, cultural identity, and social norms.

Spanning a wide range of economic, social, and political contexts, from highly developed nations like Japan to less affluent countries such as Bangladesh, the Asia-Pacific region is one of the most varied in the world. Despite significant poverty in some areas, with rates reaching up to 40%, many countries are experiencing rapid economic growth and social progress. Nonetheless, the urgent need for inclusive, equitable, and sustainable development that respects environmental limits remains paramount.

The rise of Asian economies and multinational corporations, combined with globalization and rapid technological advancements, is transforming labour markets, trade, and social structures within and beyond the region. As highlighted by the Asian Development Bank, the 21st century is increasingly recognized as "the Century of Asia Pacific", marking a pivotal shift in global focus towards this vibrant area.

This series contributes meaningfully to that transformation by providing a platform for dialogue, research dissemination, and scholarly exchange. It brings together insights from educational researchers, policymakers, and practitioners worldwide, fostering a richer understanding of educational innovation, reform, and systemic challenges unique to the Asia-Pacific context.

We warmly welcome readers interested in contributing as authors, co-authors, or editors for future volumes to reach out to the series editor directly or via Springer. We are committed to supporting prospective contributors throughout the manuscript development process to help produce high-quality publications.

June 2025

School of Education
University of Tasmania,
Hobart, Australia

Rupert Maclean AO

Faculty of Education
Monash University,
Melbourne, Australia

Philip Wing Keung Chan

Contents

List of Abbreviations

BoH	Branches of Hope
BoR	Bill of Rights Ordinance
CEDEFOP	European Centre for the Development of Vocational Training
CRAC	Central Reserve Allocation Committee
EE	Environmental Education
EfS	Education for Sustainability
ESD	Education for Sustainable Development
GGS	Generic Green Skills
GMES	Global Monitoring and Evaluation Survey
HK	Hong Kong
HKSARG	Hong Kong Special Administrative Region Government
ILO	International Labour Organization
Immd	Immigration Department
ISS	International Social Service
MSW	Municipal Solid Waste
NGO	Non-Governmental Organization
OECD	Organization for Economic Co-Operation and Development
Q.	Question
RAS	Refugees and Asylum Seekers
RE	Renewable Energy
R&D	Research and Development
SAR	Special Administrative Region
SD	Sustainable Development
SDGs	Sustainable Development Goals
SE	Social Enterprise
SMEs	Small and medium-sized enterprises
SUD	Sustainable Urban Development
TVET	Technical and Vocational Education and Training
UN	United Nations
UNHCR	The United Nations Higher Commissioner for Refugees

List of Figures

List of Tables

List of Images

Chapter 1
Introduction: Refugees and Asylum Seekers in the Context of Sustainable Urban Development

Abstract The purpose of this book is to outline the circumstances necessary to guarantee that refugees and asylum seekers (RAS) are included in sustainable urban development (SUD) processes, as well as the possible role that they may play in this green transformation. The authors report the results of a study in the Hong Kong Special Administrative Region (SAR). The book makes recommendations about how refugees and asylum seekers could enhance both their lives and urban sustainable development. It explores how formal and informal institutions may affect this potential contribution and offers strategies for integrating migrants into SUD initiatives at the theoretical and practical levels.

Particularly in this chapter, several entry points for the study are established to introduce the background of the study to be discussed in the following chapters as well as to establish its importance. Sustainable urban development and efforts of the HK government in that regard, the urgency of addressing issues of refugees and asylum seekers in the international and HK contexts, and the importance of skills development for this demographic within the context of SUD are discussed in this chapter.

Keywords Sustainable urban development in Hong Kong · Refugees and asylum seekers in Hong Kong · Generic green skills for refugees

1.1 Introduction

This book intends to identify the potential role of refugees and asylum seekers (RAS)[1] in sustainable urban development (SUD) and the conditions required to ensure their inclusion in these processes. Although the book refers to the

[1] While refugees have been legally granted a status, asylum seekers are people waiting for international protection from an asylum country.

A. T. Saral, M. Pavlova, *Education for Refugees and Sustainable Green Urban Development*, Education in the Asia-Pacific Region: Issues, Concerns and Prospects 78, https://doi.org/10.1007/978-981-95-1680-3_1

international context, the focus is on Hong Kong SAR, as the study results presented here were completed by the authors in that city.

Considering the difficulties many countries face in implementing sustainable development (SD) agendas, active participation of all groups in society is required. Inclusive, sustainable development carries a double positivity—it benefits individuals and positively impacts the societies they live in. However, most of the time, different marginalized groups lack intentions or opportunities to contribute. Therefore, this book examines the challenges of RAS's involvement in the greening process and suggests some solutions.

This chapter sets up the background of the book by introducing several entry points for the discussion presented in the following chapters, such as sustainable urban development and efforts of the HK government in that regard, the urgency of addressing issues of refugees and asylum seekers in the international and HK contexts, and the importance of skills development for this particular demographic within the context of SUD.

In terms of urbanization and greening, a large proportion of literature focuses on the importance and outcomes of greening at the governmental level and the perspectives, attitudes, and environmental values of citizens at the societal level. However, studies on refugees' and asylum seekers' experiences that are related to SUD is very limited. It has been argued that green skills can be fundamental for developing a society that supports sustainability (especially if skills development opportunities are offered, recognized, and reflected in employment schemes) (e.g., Pavlova, 2023). however, how this could influence the lives of refugees and asylum seekers remains unanswered in international studies.

In terms of green skills development required for green transition[2] and suggested by the studies (e.g., CEDEFOP, 2010; Strietska-Ilina et al., 2011), generic green skills have been identified as those required for all members of the labour force to support SDG achievement. Further studies (e.g., Pavlova, 2017, 2018) conceptualized generic green skills[3] and demonstrate its importance because these skills increase people's capacity to address environmental issues in their daily lives and work. However, there is a lack of research that examine the development of generic green skills among refugees and asylum seekers internationally. Similarly, the local literature concerning refugees and asylum seekers (RAS) in Hong Kong focuses on three areas: refugee law, asylum policies, and the lived experiences of refugees and asylum seekers (Choy & Shi, 2021). To our knowledge, there is no local study on RAS green skills development in Hong Kong, hence a gap in the literature.

[2] Green transition is the process of designing and implementing reforms to support moving away from a fossil-fuel-based economy to a renewable and low-carbon model for environmental sustainability.

[3] Generic green skills are the core skills required for shifting to a green economy. The process involves a willingness to learn about sustainability, realizing challenges before achieving sustainability, as well as capturing opportunities brought by the greening of production and consumption manners (Pavlova, 2017). For a fuller discussion, see Chap. 3.

Our passion for achieving sustainable cities with just and equally distributed resources, inclusive societies, and environmentally friendly practices (all of which contribute to SUD), including identified gaps in knowledge, led to the formulation of the focus and rationale for the research. The study presented in this book focuses on improving refugees' and asylum seekers' well-being (conceptualized in Chap. 2) by developing their capacity to think and act pro-environmentally, which is regarded as generic green skills. In this book, we argue that the development of generic green skills (GGS) can enable refugees and asylum seekers to become contributing members of society, engage in meaningful activities, express their talents, appreciate nature, and possess a sense of normalcy. Thus, the main focus of the book is based on the following question:

> To what extent does the development of generic green skills influence the well-being of refugees and their potential contribution to sustainable urban development?

This chapter focuses on refugees' position in the sustainable urban development agenda. Although most of the issues discussed in the book are relevant to different contexts, the book focuses on Hong Kong, as it presents a study conducted with RAS in this city.

1.2 Sustainable Urban Development

Currently, half the global population lives in urban areas. Estimates suggest this percentage will rise to 68% by 2050 (UN DESA, 2018). Cities attract and retain more people due to:

- the availability of jobs, better educational and health services;
- their provision of good infrastructure, connectivity, and reduced cost of doing business;
- the good quality of life and a clean, affordable, and livable environment for people and their businesses (The World Bank, 2009).

However, increased population means increased demand for housing, jobs, education and health services, food, energy, transportation, etc. To supply this demand, the extraction of raw materials from the environment increases (-especially in a linear economy where the item is produced, sold, used, and thrown away), and this type of consumption and production pattern creates pressure not only for the environmental wellbeing but also for all living in it because it contributes negatively to climate change, water and atmospheric pollution, deforestation, and loss of biodiversity, eventually making the Earth less hospitable to humans and other beings. Therefore, sustainable urban development necessitates redirecting economies into restorative rather than exploitative paths that are not reliant on the ever-growing consumption of material products, long-distance trade, and the replacement of local businesses by local branches of multinational corporations. Such economic repositioning has attested to improving sustainable development's economic, social, and environmental aspects (Pearce & Barbier, 2000).

Sustainable Urban Development (SUD) is a process for improvement, progress, and positive change in life's social and economic dimensions and the preservation of environmental resources (United Nations Environment Programme, 2011). SUD stresses urban growth through the effective use of resources, movement of people and goods, and physical infrastructure. Ensuring the sustainability of urban development requires consideration of demands (as mentioned above, such as transportation, housing, energy systems, and access to quality education and health services) and supply to fulfill the needs of an increasing population in the long term (UN DESA, 2018). In this regard, the UN Habitat-I conference in 1976 arguably marked the government's first recognition of the need for sustainable human settlements and urbanization (UN-Habitat, 2017). It is also about when SD thinking (concerns of economic growth and human-nature relationship) was brought to the front line of the international agenda. SD, as a concept, came to be defined as "the ability to meet the needs of the present without compromising the ability of future generations to meet their own needs," following the Brundtland Report (Brundtland Commission, 1987). Further developments in this area are closely linked to the UN member states' adoption of sustainable development goals (SDGs) in 2015; these provide blueprints and strategies to develop and support financially stable economies, inclusive societies, and sustainable environments.

In terms of economic development, the greening of the economy has been introduced as the model for growth, and many governments have made commitments to greening their economies, which involves practices relating to production processes and consumption patterns based on environmental concerns (Asian Development Bank, 2012). Greening of the economy is designed to ensure the efficient use of resources, integrate low-carbon practices into businesses, implement policy changes, and utilize innovation and technologies. The reason behind this shift towards greening relates to recognizing the damage caused by rapid economic growth caused by adopting traditional financial models, which has been accompanied by massive energy consumption, pollution, and carbon dioxide emission, as well as resource extraction. The model and its outcomes are predicted to be unsustainable unless issues related to society, industry, and the environment are considered together (Asian Development Bank, 2012).

Shifts toward a greener economy require a change in knowledge, skills, and technology (Bretschger, 2015) due to the creation of new jobs and changes in the nature of current jobs to support greening. Pavlova (2017, 2018) argues that green skills are vital to address the needs of a greener economy because the process of greening calls for an understanding and developing the capacity for environmentally sustainable practices, such as reducing waste and carbon emissions, using renewable energy, promoting energy efficiency, and adopting a circular system within industries. Hence, applying greener practices in daily and industrial life can influence sustainable development within urban areas (UN DESA, 2018).

The rapid growth in terms of supply and demand in urban areas is expected to transform cities into concentrated economic, social, and cultural integration hubs.

Considering that the urban population will continue to grow in the future and new cities will be formed, economic growth is expected not to be sustained by currently available environmental resources (UN DESA, 2018). Additional pressures caused by climate change and urbanization in the Asia Pacific region, particularly, have been rapid and challenging, as the area is home to the world's most polluted and unhealthy cities. Therefore, sustainable urban development is critical in providing a healthy living environment and ensuring the well-being of its inhabitants.

Regarding social development, a report by the UN-Habitat suggests that the Asia Pacific region will be impacted very heavily by climate change, and it will disproportionately affect the poor and marginalized (UN-Habitat, 2015). This situation indicates multiple dimensions to which environmental problems may be related, such as access to specific resources (i.e., housing, employment, education) and legal frameworks that may or may not guarantee marginalized groups' rights to access them, hence the need for empowerment (addressed in Chap. 3). The emphasis on people's well-being in the context of sustainable urban development has been recognized by several frameworks and action plans, such as the seventh Environment Action Plan by the European Union (European Parliament & Council of the European Union, 2013; Decision of the European Parliament and the Council on a General Union Environment Action Programme to 2020 'Living well, within the limits of our planet' (United Nations, 2012); and the OECD report on Global Wellbeing since 1820 (OECD, 2014).

From an ecological perspective, rapid and dense urbanization causes harm to the environment because cities account for 60–80% of the total energy consumption and 75% of carbon emissions globally. At the same time, they only occupy 3% of the Earth's land (United Nations). Although cities bring people and goods together, which in turn facilitates the flow of ideas, cultures, and diverse socio-economic activities (United Nations Environment Programme, 2011), they are also the places where international commitments and constraints stemming from supply and demand can come into conflict. Economic, social, and environmental challenges can be different for different groups of people; therefore, governing bodies need to implement diverse policies and practices to address their differing needs in supporting SUD and achieving a better life in cities for everyone.

Issues related to the greening of economies and sustainable development of cities have become priorities for governments and governing bodies of these cities and have been attracting increased attention from researchers. Davies and Mullin's (2011) study highlights the importance of sustainable innovation and how enterprises can be spaces for that and suggests measuring and communicating sustainability within economic spheres. Harper-Anderson (2012) portrays how women of color in the United States have been deprived of opportunities and left out. However, the policies were primarily designed for environmental protection and economic recovery, indicating the importance of resource distribution and inequality for specific demographics and areas. Wanner (2014) argues that despite the claimed 'win-win' scenario in greening economies, the dominant discourse is subsumed by capitalist hegemony and covered under 'neo-liberalization of nature,' akin to marketization, privatization, and commodification. Therefore, greening the economy and its relevant strategies are argued to be a response to the social and political dimensions of sustainability and social and international justice issues. Hence, working towards SUD involves changing how we produce (industry) and consume (behavior) and a constructive and inclusive approach for all to contribute to this growth. To support this, local governing authorities must intervene more directly in 'greening' companies, businesses, and citizens and stimulate behavioral changes toward sustainable lifestyles (While, Jonas, & Gibbs, 2004).

1.2.1 Sustainable Urban Development in Hong Kong

Historically, urbanization in Hong Kong related to the arrival of almost half a million people from Mainland China, who were considered 'refugees' over a short period around the 1950s (Hambro, 1957). Urbanization included the construction of housing estates and industrial buildings that needed more environmental considerations during this period to cater to this influx of people's housing and financial needs. (Dwyer, 1971; Prescott, 1971). Therefore, nowadays, when Hong Kong attempts to balance sustainability and urbanization, the government faces issues of poor air and water quality in addition to the pollution of marine resources, increased solid waste, and severe noise pollution (Walker & Hills, 2014). For example, contrary to casual assumptions about pollution and the city's proximity to the manufacturing centers of Mainland China, 53% of air pollution is generated locally in Hong Kong due to the historical poor planning of airflow between high-rise buildings. On the other hand, water quality has been decreasing due to a lack of adequate wastewater treatment and open garbage disposal systems. Furthermore, the demand for better housing conditions still needs to be solved. More than 170,000 people live in subdivided flats in Hong Kong (UN Habitat, 2015), including RAS.

The Hong Kong SAR Government is trying to respond to these environmental problems through different means: by involving government bodies, community, and private sectors in technological, economic, and scientific investments/developments. The aim is to bring about noticeable improvements in urban areas through quality and green-concerned landscape design, enhancing existing green areas via tree management, and creating harmony between nature and people by planning and developing public works projects (GovHK, 2022). Outlined below are some of the initiatives the HK government has put in place concerning different environmental issues that are influential in sustainable urban development:

- ***Energy efficiency:*** There are no indigenous energy resources in Hong Kong. Therefore, energy supplies are derived almost entirely from external sources. Energy is either imported directly (as in the case of oil and coal products) or produced through intermediate transformation processes using imported fuel inputs (as in the case of electricity and town gas). According to government sources, energy consumption increased by 9.8% between 2005 and 2015, accompanied by an increased population and economic and residential activities (Electrical and Mechanical Services Department, 2021). The government has introduced some initiatives to tackle this and reduce energy waste in different sectors, namely electronics, transportation, and accommodation. Initiatives include the introduction of energy labels for electronic appliances, replacing diesel buses with LPG-powered or electric/hybrid buses, and the introduction of tax incentives for energy-efficient building installations, including lighting, electronic, air conditioning, and escalator installations (HKSARG, 2018; EMSD, 2021a, 2021b; EMSD, 2020; EMSD, 2021a, 2021b). After the increased attention of HKSARG's investment in energy and environment, a new survey in the city revealed that the energy and environment sector is the most attractive sector to work for (Ranstad, 2023)

- ***Renewable Energy:*** In 2018, less than 1% of Hong Kong's total energy consumption was generated through renewable energy sources. However, there is a realizable potential for 3–4% of renewable energy production through solar, wind, and waste-to-energy systems based on mature and available technologies until 2030 (GovHK, 2021a, 2021b). Although Hong Kong has no indigenous fuel sources or the physical conditions required for renewable energy, some investments have been made. For example, expanding the electric vehicle charging network (Environment Bureau, 2011) and introducing a Feed-in Tariffs (FiT) scheme, which encourages the private sector to participate in small-scale distributed renewable energy (RE) generation by installing RE systems at their premises (GovHK, 2021a, 2021b), such as O-Park1 (O-Park, 2020). In Hong Kong, O-Park1 is the first organic waste recovery center that turns food waste into biogas (a renewable energy source). The government has been calling for expressions of interest from commercial and industrial sector players to sponsor activities for this initiative. Partnering with such stakeholders would benefit the effective running of the plant and attract more customers to the organization's premises due to its corporate social/environmental responsibility. In addition to that, as RE is an emerging sector, a workforce with the necessary skills is required to work in RE plants and transmission platforms. Therefore, the programs offered and their accessibility to the community are of concern to meet the need.
- ***Waste management:*** Although statistics show that municipal solid waste (MSW) in Hong Kong, which consists of domestic, commercial, and industrial waste, decreased in 2019 compared to 2018, the recovery rate decreased, which has resulted in increased disposal at landfills (Environmental Protection Department, 2020). The major component of MSW in landfills is food waste, followed by paper, plastics, and others (such as wood, textiles, and bulky items). WEEE-Park is a treatment and recycling facility in the Tuen Mun area that accepts electrical and electronic equipment (WEEE-Park, 2021). This eco-park implements advanced technologies to turn waste into valuable secondary raw materials through detoxification, dismantling, and recycling processes. The processes result in waste becoming resources again as they are reused/repurposed. In addi-

tion, the government is collaborating with different industries, such as the hotel industry, to find green solutions for waste. With the support of the Hong Kong Hotels Association, the Glass Container Recycling Program was introduced in 2008. This program turns collected waste glass containers from hotels into paving blocks (Environmental Protection Department, 2014). However, HK's environmental mitigation indicators, such as waste management and carbon emission, are still inadequate (Arcadis, 2018), indicating the need for education and training to raise public awareness.

- ***Air quality:*** The main reasons for poor air quality in Hong Kong are proximity to the industrialized area in the Pearl Delta Region, closely spaced high-rise buildings on both sides of roads, traffic congestion due to the increased number of vehicles, and emissions from shipment activities. The introduction of new cycling networks in the city, such as the Tuen Mun to Ma On Shan route, as well as the use of hybrid buses combined with the construction of vertical gardening, green roofs, and tree planting in urban design and construction sectors, are examples of initiatives designed to improve air quality in the city (Environment Bureau, 2013).
- ***Water quality:*** Government sources show that the growing population and commercial & industrial activities have increased Hong Kong's freshwater consumption over the past decade (Research Office—LegCo, 2021). Without natural lakes, large rivers, or underground water, Hong Kong has mainly depended on imports from the Dongjiang River since 1965 to meet local demand. In collaboration with Guangdong authorities, one project is being implemented to secure a stable and adequate water supply to meet the needs and diversify the water supply in Guangdong. This project will divert water from the Xijiang River (which has a bigger flow than the Dongjiang River) in the western Pearl River Delta to the eastern Delta region cities, which include Guangzhou Nansha, Dongguan, and Shenzhen. Once complete, the project is expected to enhance the backup water supply to Hong Kong (Research Office-LegCo, 2021). In addition to providing quality freshwater, treating wastewater at sewage plants for safe use is essential for water supply worldwide. For this reason, the government established a sludge-treatment facility called T-Park (T-Park, 2021). The facility generates energy from sludge, and residue from this process is turned into fertilizers used in the facility's landscape design.
- ***Noise:*** As Hong Kong's economy and population grew, the noise from construction, transportation, commercial and industrial sources increased, too. Such noise can lead to mental stress and hearing loss and influence daily activities such as completing homework, watching television, talking on the telephone, and sleeping. Noise can add to the daily stress of people, disturbing their well-being. To control the problem and regulate noise emissions, the government has passed a Noise Control Ordinance. Some examples of the mitigation of noise in Hong Kong include the provision of soundproofing systems to schools near major traffic roads, the installation of noise barriers by the side of the streets, and encouraging the use of quieter construction equipment (GovHK, 2021a, 2021b; Environmental Protection Department, 2014, 2020).

The above initiatives indicate the importance the government places on greening and the city's sustainable development by adopting initiatives, introducing ordinances, and collaborating with multiple stakeholders related to environmental issues. However, SD, as stipulated in SDGs, should be supplemented by measures that are focused on being inclusive and building the capacity of all members of society to ensure:

- the alleviation of poverty and reduction of deprivation;
- the provision of equal access to opportunities or development of capabilities; and
- quality of work and reduced inequality in outcomes (McGrath & Powell, 2016).

Thus, many studies have questioned government measures towards SUD. In Hong Kong, Harris (2012) argues that geographic features are destroyed to provide housing and economic progress by extensively leveling the mountains and islands and reclaiming the sea. While doing that, the entire local ecosystem was destroyed. Looking at SD in Hong Kong between 1995 and 2016, Shi et al. (2020) present that although the Human Development Index (HDI) rose during this period, the per capita ecological footprint of Hong Kong increased. The consumption of fossil energy had a crucial impact on the city's ecological footprint, whereas the biocapacity of Hong Kong declined gradually (Shi et al., 2020). From an SDG perspective, it is important to eradicate poverty and enable people to have dignified and comfortable lives in harmony with nature. Although there are improvements, the poverty rate still stands at 7.9% even after all governmental interventions are considered (Census and Statistics Department, 2021). Looking at poverty and well-being among the elderly in the city, Fong and Law (2022) indicated that nearly four in every ten elderly people were living in poverty.

Similar criticism is evident across international studies. Moloney and Strengers (2014) look into participation in 'easy' sustainable actions in Australian households and find that there are many consumption habits ingrained in everyday life that focus just the actions that do not lead to social change, especially when our lives are becoming more resource extensive due to external factors such as transportation, housing, and infrastructure (Moloney & Strengers, 2014). Hence, suggesting a whole approach including policies, strategies and practices. Malier (2019) analyses how poor households are responsibilized (shift of caring for others from state to individuals—in this case, poor households -) for sustainability through programs, and their lifestyles are framed as 'eco-friendly' and moralized in national programs in France. Yet, this demonstration did not come from disadvantaged groups' environmentalism. In one case, environmental responsibilization was combined with responsibilization against fuel poverty (self-care).

We strongly believe that all individuals should contribute to the greening agenda despite their position in the social hierarchy; however, in the case of disadvantaged populations such as refugees and asylum seekers, their contribution should also benefit them in ways they can feel to maintain the behavior. Supporting this notion, Angelo (2019) argues that initiatives concerning urban greening often address symptoms, not necessarily the root cause, and hence can fail to motivate society intrinsically. Therefore, we should also understand refugees and asylum seekers situation when trying to involve them as they are often neglected in the process (e.g., Malier, 2019; Angelo, 2019),

1.3 Refugees and Asylum Seekers (RAS)

Seeking refuge is a multi-dimensional issue involving crossing borders and interactions with state-led institutions. By 2024, 43.4 million refugees sought international protection (UNHCR, 2024). Most refugees live in urban areas, hoping to gain better access to resources, including education, employment, and health services. The United Nations High Commissioner for Refugees (UNHCR) emphasizes the importance of collaborative work for skills development and jobs for displaced populations, which can help stimulate economic growth and development within host communities (UNHCR, 2019). This *focus on engaging refugees* enhances universal access to human rights and consolidates the attainment of various sustainable development goals (SDGs), such as access to inclusive and quality lifelong education (SDG 4), reduced inequality (SDG 10), sustaining decent work and economic growth (SDG 8), as well as establishing sustainable cities and communities (SDG 11), all of which contribute to good health and wellbeing for human beings (SDG 3).

Refugees and asylum seekers (RAS) have often been excluded from sustainability and wellbeing-related frameworks. The fact that inequality and exclusion remain a persistent trend in urban areas (UN-Habitat, 2020) requires subnational and local governments to fulfill their roles in creating opportunities for dialogue, including age- and gender-responsive approaches for inclusive and sustainable growth, with particular attention to potential contributions from all segments of the society, including RAS. However, this distinct group might need specific conditions for engagement (UN-Habitat, 2017).

However, political discourses and increasing research on migration and solution-oriented approaches for integrating RAS internationally that highlighted the urgency of addressing the needs of refugees and their inclusion in new societies, education, and employment have yet to be mainstreamed in Hong Kong. As Brown (2008) argues, in the short term, forced migration due to climate change can influence the achievement of SDGs. In the long term, climate change-forced migration (such as migration due to sea-level rises) can roll back the progress that has been achieved so far because it interrupts the provision of education and health-related services, both of which relate to several SDGs such as universal primary education (SDG 2) and reducing child and maternal mortality, as well as combating HIV/AIDS, malaria and other diseases (SDG 4 and 5) (Brown, 2008). Therefore, regardless of whether migration is caused by climate change or other issues, the capacity of the group forced to leave their origin needs to be improved if these people are to contribute to the sustainability efforts of their host states because once they enter a new setting, they have already experienced a loss of resources and feel powerless.

1.3.1 The Urgency of the Issue

In terms of tackling issues related to injustice (Loper, 2010), inequality (Oxfam, 2018), poverty (Ng et al., 2021), education (Tsang et al., 2020), and participation in urbanization (Fong, 2001), the HKSARG is under pressure to provide more equitable access to resources and education for all groups to improve human potential. Different researchers have considered various perspectives on this topic. This inequality is particularly relevant in the context of equitable allocations to refugees and asylum seekers. The current situation concerning refugees and asylum seekers puts the HKSARG under critique. When analyzing asylum seekers' daily experience of adaptation, suffering, and bargaining to survive in Hong Kong, Vecchio (2016) found that social borders defined by legal means prevent asylum seekers from exerting their agency. So, their lives and futures are controlled. As a consequence, they indicate they are "unhappy". Financial problems and a lack of lawful opportunities to engage in the formal economy or education profoundly impact their decision-making and livelihoods. The refugee and asylum seekers' experience under HK policies is variously described as "in transit" (Thomas, 2000), "living in limbo" (Booth, 2020), or that they are "undesirable guests" (Vecchio, 2016).

In terms of discourse in Hong Kong, Kennedy et al. (2019) observe the following:

1. The government's discourse is described as instrumental at best and anti-humanitarian at worst because they emphasize being strict about not granting refugee status. However, it does comply with court rulings concerning torture or persecution claimants.
2. The discourse of NGOs and lawyers, as a community that focuses on asylum seekers' human rights, is progressive or liberal because the community often conflicts with the government and pushes for proactive and humanitarian legal support.
3. The discourse of the society contains demeaning attitudes whereby refugees and asylum seekers are viewed as illegal immigrants, overstayers, and criminals. These attitudes are sometimes reflected in the media or on the Security Panel of the Legislative Council. Therefore, society's discourse is described as populist or racist.

Negative perceptions towards refugees at the public and institutional level partly stem from Hong Kong's historical experience of refugee influxes—such as the Chinese immigrants, the Vietnamese 'boat people,' and recent arrivals—and are etched into the city's psyche. The majority of people still use the terms 'illegal immigrants' and 'asylum seekers' interchangeably (McKenzie 2009 in Vecchio, 2016). The media also has a role in characterizing the group, manipulating people's attitudes, and creating a panic environment to gain high circulation rates (Trier, 1994). A local study looked into the print media (news articles and magazines) and considered how it portrayed refugees and asylum seekers between 2015 and 2016 in Hong Kong. They are represented as criminals and fake (Ng et al., 2018). Although media constructions cannot be regarded as solely responsible for creating negative discourses about refugees and asylum seekers, when the legislative council (the city's elected administrative body) reiterates such images, a xenophobic and hostile attitude emerges in society (Justice Centre HK, 2016). Considering that sustainability encompasses social, economic, and environmental aspects, HK's attempts to achieve sustainable urban development would be more robust if they created a harmonious society where they look after refugees' well-being as locals.

Thus, to successfully implement the SUD agenda, the HKSARG needs to identify what part all groups in society play in this process and the impact of providing opportunities to alleviate the conditions of refugees. Attaining a sustainable urban life relies on a positive and nurturing relationship with nature, not simply a consideration of raw materials, water, or other ecosystems nature provides. Nature can increase one's capacity to care, reason, love, create, and understand more about sustaining a good life (Bratman et al., 2015). Therefore, it enables sympathizing with those in need and creating a conscience to reduce inequalities. Kellert (2012) argues that with increased urbanization, human relationships with nature have become adversarial, and this situation endangers not only environmental life but also social lives. However, one can repair this situation by realizing how much health and well-being relies on being a part of nature rather than being apart. Discovering where one fits in this scheme requires a much deeper and fuller understanding of what nature can provide. Formal and non-formal education and learning platforms can be avenues to discuss that (see Chap. 3).

Therefore, there is a need to raise awareness, encourage action regarding environment-related challenges, and involve the community at all levels, including refugees and asylum seekers. This would not only benefit the city's and community's sustainability (SDG 11), but it would also promote decent work for all and economic growth without harming the environment (SDG 8), hence increasing the well-being of its residents (SDG 3) (Milfont & Duckitt, 2010; Stone et al., 1995; Erzengin & Cetin Teke, 2013; Bogner, 2018). While supporting this notion, Colantonio and Dixon (2011) look into the sustainability of societies and communities at the basic level (e.g., shelter, food, clean water, and employment) and the operational level (capacity building, skills development, equity, distribution of, and access to, resources, etc.). They state that societies with lower levels of disparity are characterized by longer life expectancies, fewer homicides and fewer crimes, stronger patterns of civic engagement, and a more robust economic vitality (cited as GVRD, 2004a in Colantonio & Dixon, 2011, p.24). The same has been confirmed by two NGOs working with refugees and asylum seekers in Hong Kong during the pilot study conducted by the researchers. NGO representatives stated that refugees and asylum seekers can gain dignity by contributing to society. To enable their potential, these groups require empowerment, which would positively contribute to the sustainability of societies, economies, and the environment.

1.4 The Overall Design, Type, and Strategy for the Research

This research aims to explore the perspectives and behaviors of refugees and asylum seekers regarding environmentally sensitive practices and issues related to their well-being. We utilized a qualitative design and interpretive approach to explain and interpret the potential of engaging RAS in sustainability and social entrepreneurship based on their perspectives (Mertler, 2019; Maxwell, 2009; Braa & Vidgen, 1999).

Another important feature of this study is its evaluative nature. We designed an intervention through workshops and then evaluated its results. We used Kirkpatrick and Kirkpatrick's (2006) four-level evaluation model. The levels indicated in this model are related to the outcomes we expected to see in our project: feedback, learning, behavior, and results (Kirkpatrick & Kirkpatrick, 2006). In this model, each level has an influence on the next.

However, as we explained elsewhere (Saral & Pavlova, 2025), our study does not follow what Kirkpatrick and Kirkpatrick (2006) suggest in terms of measurement for change, such as creating a control and experiment group, as well as collecting data for three to 6 months after the conclusion of training for a couple of reasons: i) human behavior involves more factors than those researchers can control in a laboratory (Payne & Payne, 2004), ii) due to time and financial constraints during data collection. Additionally, current policies and practices in Hong Kong do not permit us to guarantee employment following participation in our workshops and/or organizational training offered by NGOs or the government, which is required for an effective outcome. Consequently, Pulichino's (2007) elucidation of Kirkpatrick and Kirkpatrick's definition of evaluation for each level served as a guide for developing our understanding of acceptable evaluation for this study. We adjusted it based on our application in the study (see Table 1.1).

The intervention strategy involved organizing workshops. These workshops introduced acute SD-related concepts in Hong Kong, covering their applications in daily life as well as problems related to each concept. Case studies, on the other

Table 1.1 Four levels of evaluation

Level	Definition	Our application in the study
Level 1—Reaction	How students react to the training	The feedback from green refugee day and motivation to participate in the study
Level 2—Learning	The extent to which students change attitudes, improve knowledge, and/or increase skills as a result of the training	How student behaviors, attitudes, knowledge, and values changed as a result of the training
Level 3—Behavior	The extent to which on-the-job behavior or performance changed and/or improved as a result of the training	The applicability of the skills in their daily lives or at home
Level 4—Results	The extent to which desired business and/or organizational results have occurred due to the training.	Possibility of the realization of these skills via a social enterprise

Source: Authors, adapted from Pulichino, 2007

hand, can be considered industry examples that demonstrate successful solutions to the problems outlined in the concept sheet. There were activities related to concepts and/or case studies designed to foster learning, as well as hands-on activities that demonstrate applicability. The materials used in this project were previously developed as part of a package for Dr. Margarita Pavlova's CRAC project (Creating an Impact: Setting up a Green Skills Hub at the EdUHK—03A68) and were deemed successful.

Regarding the content of the workshops, the concepts for the workshops were selected based on three considerations: (i) their relevance to the Hong Kong context and needs, (ii) their applicability at home, and (iii) their transferability to social enterprises. It is essential that refugees recognize the benefits of participating in the workshops and sustainability programs. The potential of generating income through social enterprises is motivating, especially for those living in financially challenging conditions. While arranging the hands-on activities, we strove to ensure that learning would be fun and adaptable to refugees' lives. We focused on the 5Rs of waste management, informed by local studies (e.g., Pavlova, 2019; Yau et al., 2018).

Workshops aimed to raise awareness about the importance of the five Rs (see Fig. 1.1) incorporated into the intervention over a six-week period. The selected topics and hands-on activities are presented in Table 1.2:

In terms of participants, this study welcomed refugee women, along with their children aged between 5 and 10 years old, through a specific NGO that worked with refugees. Although the participants cannot be regarded as representative of the entire RAS population in Hong Kong, mothers and children were significant for the study because women and children are among the most vulnerable groups in terms of migration. Additionally, women comprise 50% of Hong Kong's asylum population (Issuu Inc., 2019).

We initially targeted 20 participants for each week's workshop. We had 8–10 refugee and asylum-seeking women, along with their children, attend each week, which amounted to around 15 people. Extending an invitation to children was

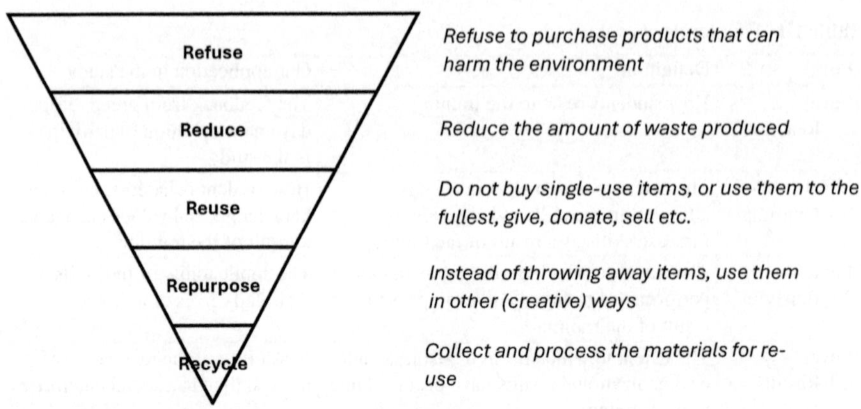

Fig. 1.1 Five Rs of Waste Management. Source: Circle Waste (2020)

Table 1.2 Weekly organization of workshops

	Name of the concept	Case study	Activities related to concept/case study	Hands-on activity (5Rs)	
Week 1	1. *Introduction of the study and activities* 2. *Collection of signed consent forms* 3. *Completion of initial survey*				
Week 2–5	Concept introduction and relevant problems about that concept (e.g., sustainable consumption, waste management, social enterprises)	A successful example addressing the issue raised in the concept (e.g., riders for health in Africa for social Enterprise concept)	An activity that fosters learning and reflection about the case study or concept introduction (e.g., group discussion on the reasons to consider or not consider sustainable consumption and how our consumption habits influence our and others' Well-being)	A hands-on, DIY-like greening activity that can be done at home or through social enterprises (such as soap recycling or turning used cooking oil into candles)	**Relevant R**
Week 6	1. *Completion of end survey* 2. *Focus group discussion* 3. *Distribution of certificates*				

decided for a couple of reasons. Children can learn specific behaviors from their parents and with them (Goldberg & Carlson, 2014), which can affect their well-being and development. Also, refugees often do not have a place to leave their children if they want to attend an event or training. Therefore, welcoming their children, unless they had to attend school, was a gesture to remove barriers to attendance. Additionally, participants were offered supermarket coupons to compensate for their transportation expenses.

Despite the relatively small sample size, participants remained engaged for more than 6 weeks, allowing for the collection of in-depth data required for a qualitative study. We employed surveys, interviews, focus group discussions, and observations. In our analysis, we classified general themes based on the identified concepts, as recommended by Naeem et al. (2023). Therefore, we performed analyses based on the themes identified in Chaps. 2 and 3 of this book.

On the other hand, the survey data was analyzed automatically using the Qualtrics platform based on the total response count per item. The platform helped visualize the answers of a small group of participants, supporting the selection of identified themes, as seen in Chaps. 4 and 5.

1.5 Significance of the Book

The significance of this book can be considered at both the theoretical and practical levels. The following areas outline the main contribution to the knowledge:

1. Addressing a research gap: The results of the study presented in the book fill the gap in the literature about the potential contribution of refugees and asylum seekers to environmentally sustainable urban development. This is the first study to examine the development of generic green skills to improve RAS' lives, specifically in Hong Kong.
2. Novelty: Currently, there is a need for more studies that focus on the relationship between refugees, sustainable urban development, and generic green skills. In the book, we present how skills development could influence this community's well-being (Chap. 2). In addition, we use the voices of refugees and asylum seekers to illustrate their beliefs and perceptions on greening, as the voices of the disadvantaged have been argued as valuable insights in academic literature (Chap. 4 & 5). We explore the opportunity structure in the city and the agency of refugees to participate in skills training and greening (Chap. 2) and discuss empowering them through the development of social entrepreneurship as part of GGS (Chap. 3). Therefore, the results of this study bring new knowledge in terms of understanding the current level of sustainability and environmental awareness among claimants in Hong Kong, as well as their level of motivation to improve their consumption and living patterns to address environmental issues and improve their well-being.
3. Design of intervention: From a methodological perspective, the study discussed in the book applies an intervention methodology for developing generic green skills for RAS' wellbeing and urban development, which has not been done before in Hong Kong-related studies. Appropriate teaching and learning methods have been designed and evaluated by applying them to work with disadvantaged communities.
4. Effectiveness of pedagogy and content for the intervention: The intervention results evaluate the effectiveness of training and have been measured by the change in knowledge, action, and attitude before and after the intervention. Therefore, Chap. 5 reviews RAS's experience in detail.

The practical contribution of this book can be viewed on three levels: personal, community, and society. On an individual level, training intends to improve the well-being of refugees and asylum seekers. The study seeks to understand whether they value this training and consider it a resource gain, as discussed in Chap. 4. On the community level, the results indicate to what extent generic green skills could empower the community to participate in social entrepreneurship after they have permission to do so. On the societal level, the discussion presented in the book seeks to demonstrate the potential of integrating refugees and asylum seekers into the generation of green solutions for a more sustainable city. Thus, recommendations to HKSARG on how sustainable urban development can be improved by integrating refugees and asylum seekers into its sustainability policy and practices are presented in Chap. 6.

Although this study does not pretend to be generalizable to all contexts, based on our working experience in different regions, we suggest that some results, including the approach for training generic green skills, will work for other settings.

1.6 Summary

Climate change and environmental challenges the world faces now require all groups of people to make an effort to facilitate change. Many governments, including Hong Kong, adopted substantial initiatives and practices at the city level for greening, including investment in recycling infrastructure, greening the landscape, and monitoring energy consumption. Hong Kong, as a large metropolis, is confronted by challenges regarding sustainability, just like many other urban areas that have become a hub for people seeking better opportunities in terms of employment, education, housing, and health services. Refugees and asylum seekers are one of the demographics seeking protection, a sense of stability, and more humanitarian living conditions. However, local governments must consider refugees' potential contributions to sustainable urban development (SUD). Hong Kong's approach to refugee policies has been criticized locally and internationally for being strict and limiting (Kennedy et al., 2016; Vecchio, 2016). Local studies about refugees in Hong Kong mainly focus on three areas: refugee law, asylum policies, and the lived experiences of refugees (Choy & Shi, 2021). Refugees' psychological and mental well-being was reported to be destitute (Wong et al., 2017) due to factors such as the long waiting periods for application processing (Lau & Gheorghiu, 2018). In addition, political and societal discourses discriminate against this group (Kennedy et al., 2019). Therefore, refugee studies related to Hong Kong have yet to focus on this group's capacity and skills development. However, some international projects have highlighted the importance of skills development for improving refugees' well-being and their contribution to SUD (UNHCR, 2019; Mucino-Sanchez, 2022; Siegfried, 2022).

The study presented in this book focuses on how Hong Kong refugees and asylum seekers can be empowered to contribute to Hong Kong's greening efforts and how they can potentially contribute to improving urban sustainable development and their lives. It discusses the impact of formal and informal institutions on this potential contribution and suggests ways of including refugees in SUD efforts at the levels of theory and practice. The following issues will be addressed in the book:

1. The perception of well-being, sustainability, and environmentally sensitive practices among refugees and asylum seekers in Hong Kong;
2. The use of a specific pedagogy to develop generic green skills for refugees and asylum seekers and its impact on the perception of sustainability and environmentally sensitive practices;

3. Access to resources for refugees and asylum seekers and the perception of the applicability of generic green skills in establishing a social enterprise as a way of empowerment and its benefit to their wellbeing.
4. Areas that the governments should consider ensuring RAS participation in SUD

The book consists of six chapters, all focusing on different threads related to refugees' lives, such as agency, opportunity structure, and training; however, this is all connected to how it influences their well-being and can be improved. The introduction and conclusion allow us to indicate the research gap and provide recommendations. The other chapters conceptualize research, present a suitable approach and pedagogy for the design of training, offer a methodological approach that can be relevant internationally, discuss findings from the mentioned study, and justify recommendations.

References

Angelo, H. (2019). Added value? Denaturalizing the "good" of urban greening. *Geography Compass, 13*(8). https://doi.org/10.1111/gec3.12459

Arcadis. (2018). *Citizen centric cities the sustainable cities index 2018*. Arcadis.

Asian Development Bank. (2012). *Policies and practices for low-carbon green growth in Asia: Highlights*. The Philippines: Asian Development Bank. Retrieved from ADB-ADBI Study on Climate Change and Green Asia: https://www.adb.org/sites/default/files/publication/29767/policies-green-growth-asia-highlights.pdf

Bogner, F. X. (2018). Environmental values (2-MEV) and appreciation of nature. *Sustainabilty, 10*(2), 350. https://doi.org/10.3390/su10020350

Booth, M. (2020). *Living in limbo: The life of an asylum seeker in Hong Kong*. (D. Zhang, Producer, & South China morning post) Retrieved September 14, 2021, from https://www.youtube.com/watch?v=DOEokI-isbY

Braa, K., & Vidgen, R. (1999). Interpretation, intervention, and reduction in the organizational laboratory: A framework for in-context information system research. *Accounting, Management and Information Technologies, 9*, 25–47. Retrieved from https://reader.elsevier.com/reader/sd/pii/S0959802298000186?token=21D670A47B1823E22D0E2A56BE57B678C94C341A34874DD6FD5FEDF73E118BA479DF516CDCCEFE985FC2093CEFD101E2&originRegion=us-east-1&originCreation=20220411044935

Bratman, G. N., Hamilton, J. P., Hahn, K. S., & Daily, G. C. (2015). Nature experience reduces rumination and subgenual prefrontal cortex activation. *Psychological and Cognitive Sciences, 112*(28), 8567–8572. Retrieved from https://www.pnas.org/doi/10.1073/pnas.1510459112

Bretschger, L. (2015). *Greening economy, Graying Society*. CER-ETH Press. https://mtec.ethz.ch/content/dam/ethz/special-interest/mtec/cer-eth/resource-econ-dam/documents/people/lbretschger/Bretschger_Greening_JUNE2015.pdf

Brown, O. (2008). *Migration and climate change*. International Organization for Migration.

Brundtland Commission. (1987). *Report of the world commission on environment and development: Our common future. World commission on environment and development*. Retrieved from https://sustainabledevelopment.un.org/content/documents/5987our-common-future.pdf.

CEDEFOP. (2010, July). *Briefing note*. Retrieved from skills for green jobs: Chrome-extension://efaidnbmnnnibpcajpcglclefindmkaj/https://www.cedefop.europa.eu/files/9024_en.pdf

Census and Statistics Department. (2021). *Hong Kong poverty situation report 2020:*. Hong Kong: HKSARG. Retrieved from https://www.censtatd.gov.hk/en/data/stat_report/product/B9XX0005/att/B9XX0005E2020AN20E0100.pdf

Choy, W. C., & Shi, X. (2021). A comprehensive review of refugee and asylum studies in contemporary Hong Kong: Law, policies, and lived experiences. *Asian Journal of Social Science, 49*(1), 38–48. https://doi.org/10.1016/j.ajss.2020.11.002

Circle Waste. (2020, September 16). *What are the 5 R's of waste management?* Retrieved from: https://www.circlewaste.co.uk/2020/09/16/what-are-the-5-rs-of-waste-management/

Colantonio, A., & Dixon, T. (2011). Social sustainability and sustainable communities: Towards a conceptual framework. In a. Colantonio, T. Dixon, & W. O. Library (Ed.), *Urban Regeneration & Social Sustainability: Best practice from European cities* (pp. 18–36). https://doi.org/10.1002/9781444329445.ch2.

Davies, A. R., & Mullin, S. J. (2011). Greening the economy: Interrogating sustainability innovations beyond the mainstream. *Journal of Economic Geography, 11*, 793–816. https://doi.org/10.1093/jeg/lbq050

Dwyer, D. J. (1971). Introduction. In D. J. Dwyer (Ed.), *Asian urbanization: A Hong Kong casebook* (pp. 1–10). Hong Kong University Press. Retrieved from https://www.jstor.org/stable/j.ctt2jc115

Electrical and Mechanical Services Department. (2021, November 15). *Energy scene of Hong Kong, China*. Retrieved from ENERGY: https://www.emsd.gov.hk/energyland/en/energy/energy_use/energy_scene.html

EMSD. (2020, December 31). *Buildings energy efficiency ordinance (cap. 610)*. Retrieved from BEEO: https://www.emsd.gov.hk/beeo/

EMSD. (2021a, October 5). *Energy efficiency registration scheme for buildings*. Retrieved from Energy Efficiency and Conservation: https://www.emsd.gov.hk/en/energy_efficiency/energy_efficiency_registration_scheme_for_building/

EMSD. (2021b, November 2). *LPG vehicle scheme*. Retrieved from Gas Safety: https://www.emsd.gov.hk/en/gas_safety/lpg_vehicle_scheme/

Environment Bureau. (2011, April). *Green Hong Kong*. Retrieved from EPD: https://www.epd.gov.hk/epd/sites/default/files/epd/english/news_events/newsletter/files/Green_HK_Apr_2011_eng.pdf

Environment Bureau. (2013). *A clean air plan for Hong Kong. Environment Bureau, in collaboration with transport and housing Bureau, Food and Health Bureau, development Bureau*. Environment Bureau. Retrieved from https://www.enb.gov.hk/en/files/New_Air_Plan_en.pdf

Environmental Protection Department. (2014, March 11). *Glass container recycling programme for the hotel sector*. Retrieved from Waste: https://www.epd.gov.hk/epd/english/environmentinhk/waste/eco_responsibility/n_gcrp.html

Environmental Protection Department. (2020, January 21). *Waste data and statistics*. Retrieved from https://www.wastereduction.gov.hk/en/assistancewizard/waste_red_sat.htm

Erzengin, O. U., & Cetin Teke, E. (2013). A study on developing an environmental behavior and attitude scale for university students. *Journal of Educational and Instructional Studies in the World, 3*(2), 49–56. Retrieved from https://www.researchgate.net/publication/320716991_A_STUDY_ON_DEVELOPING_AN_ENVIRONMENTAL_BEHAVIOR_AND_ATTITUDE_SCALE_FOR_UNIVERSITY_STUDENTS/link/59f75639458515547c24c123/download

Fong, B. Y., & Law, V. T. (2022). *Sustainable development goal 3: Health and Well-being of aging in Hong Kong*. Routledge. Retrieved from https://ebookcentral.proquest.com/lib/hkive/reader.action?docID=7245593&query=&ppg=5

Fong, G. K. (2001). *Public participation in Hong Kong: Case studies in community urban design (China)*. Chinese University of Hong Kong. Retrieved from http://pqdd.sinica.edu.tw/doc/3025925

Goldberg, J. S., & Carlson, M. J. (2014). Parents' relationship quality and children's behavior in stable married and cohabiting families. *Journal of Marriage and Family, 76*(4), 762–777. https://doi.org/10.1111/jomf.12120

GovHK. (2021a, November). *Installation of renewable energy systems*. Retrieved from Renewable Energy: https://www.gov.hk/en/residents/environment/renewable/installation.htm.

GovHK. (2021b, March). *Key government renewable energy projects*. Retrieved from https://www.gov.hk/en/residents/environment/renewable/projects.htm

GovHK. (2022, April). *Greening Hong Kong*. Retrieved from GovHK: https://www.gov.hk/en/residents/environment/sustainable/greening/greening.htm

Hambro, E. (1957). Chinese Refugees in Hong Kong. *The Phylon Quarterly, 18*(1), 69–81. Retrieved from https://www.jstor.org/stable/273076?seq=3#metadata_info_tab_contents

Harris, P. G. (2012). *Environmental policy and sustainable development in China: Hong Kong in global context*. Policy Press.

Harper-Anderson, E. (2012). Exploring what greening the economy means for African American workers, entrepreneurs, and communities. *Economic Development Quarterly, 26*(2), 162–177. Retrieved from https://journals.sagepub.com/doi/full/10.1177/0891242411431957

HKSARG. (2018, March 1). *Enhanced tax incentive to promote renewable energy and building energy efficiency*. Retrieved from Press Release: https://www.info.gov.hk/gia/general/201803/01/P2018030100661.htm

Issuu Inc. (2019, December 1). *Refugees*. Retrieved from Issuu: https://issuu.com/arianalife/docs/ariana04/72?fbclid=IwAR0vpYVrfIVtYGL7nc53MH_aYnN3XWJTK0lNQStlSJzuV-iCGYHYkhKFdT0

Justice Centre HK. (2016, April 11). *Joint statement*. Retrieved from Justice Centre Hong Kong: https://www.justicecentre.org.hk/jointstatement-eng/

Kellert, S. R. (2012). *Birthright: People and nature in the modern world*. Yale University Press. Retrieved from https://www.jstor.org/stable/j.ctt32bq08

Kennedy, K. J., Wong, K. L., & Ng, H.-Y. (2019). Being asylum seekers in Hong Kong: Institutions, action and power. *Asian Journal of Social Science, 47*, 534–565. https://doi.org/10.1163/15685314-04704006

Kennedy, K., Wong, K. L., & Ng, H. Y. (2016). *Towards a refugee policy for an inclusive Hong Kong: Enhancing the status of China's International City. The education University of Hong Kong, Centre for Governance and Citizenship*. Central Policy Unit of the HKSAR Government. Retrieved from https://www.pico.gov.hk/doc/en/research_report(PDF)/2014_A5_010_15B_Final_Report_Prof_Kennedy.pdf

Kirkpatrick, D. L., & Kirkpatrick, J. D. (2006). *Evaluating training programs: The four levels*. Berrett-Koehler Publishers, Incorporated.

Lau, P. Y., & Gheorghiu, I. (2018). Vanishing selves under Hong Kong's unified screening mechanism. *Cultural Diversity in China, 3*(1) Retrieved from. https://doi.org/10.1515/cdc-2018-0003/html

Loper, K. (2010). Human rights, non-Refoulement and the protection of refugees in Hong Kong. *International Journal of Refugee Law, 22*(3), 404–439. Retrieved from https://hub.hku.hk/bitstream/10722/92989/2/Content.pdf

Malier, H. (2019, December). Greening the poor: The trap of moralization. *The British Journal of Sociology, 70*(5), 1661–1680. https://doi.org/10.1111/1468-4446.12672

Maxwell, A. J. (2009). Chapter 3: Conceptual framework: What do you think is going on? In A. J. Maxwell (Ed.), *Qualitative research design: An interactive approach* (pp. 39–72). SAGE Publications.

McGrath, S., & Powell, L. (2016). Skills for sustainable development: Transforming vocational education and training beyond 2015. *International Journal of Educational Development, 50*, 12–19.

Mertler, C. A. (2019). *Introduction to educational research*. SAGE.

Milfont, T. L., & Duckitt, J. (2010, March). The environmental attitudes inventory: A valid and reliable measure to assess the structure of environmental attitudes. *Journal of Environmental Psychology, 30*(1), 80–94. https://doi.org/10.1016/j.jenvp.2009.09.001

Moloney, S., & Strengers, Y. (2014). 'Going green'?: The limitations of behaviour change Programmes as a policy response to escalating resource consumption. *Environmental Policy and Governance, 24*(2), 94–107. https://doi.org/10.1002/eet.1642

Mucino-Sanchez, A. (2022, November 22). *Somali bantu refugees offer new roots and bountiful harvest in Maine*. Retrieved from UNHCR USA: https://www.unhcr.org/us/news/stories/somali-bantu-refugees-offer-new-roots-and-bountiful-harvest-maine?_gl=1*1f8xxn7*_rup_ga*MTIzMTUxNzYwNy4xNzM3MDE3MzA2*_rup_ga_EVDQTJ4LMY*MTczNzAxNzMwNi4xLjEuMTczNzAxNzk4NS42MC4wLjA.

Naeem, M., Ozuem, W., Howell, K., & Ranfagni, S. (2023). A step-by-step process of thematic analysis to develop a conceptual model in qualitative research. *International Journal of Qualitative Methods, 22*. https://doi.org/10.1177/16094069231205789

Ng, I., Choi, S. F., & Chan, A. L. (2018). Framing the issue of asylum seekers and refugees for tougher refugee policy–a study of the media's portrayal in post-colonial Hong Kong. *Journal of International Migration and Integration*, 1–25.

Ng, M., Lau, Y., Chen, H., & He, S. (2021). Dual land regime, income inequalities and multifaceted socio-economic and spatial segregation in Hong Kong. *Urban Socio-Economic Segregation and Income Inequality, 113–133*. https://doi.org/10.1007/978-3-030-64569-4_6

OECD. (2014). In van Zanden, J. l., et al. (eds.), *How Was Life?: Global Well-being since 1820*, OECD Publishing. https://doi.org/10.1787/9789264214262-en

O-Park. (2020). *O-Park is...* Retrieved from https://www.opark.gov.hk/en/index.php

Oxfam. (2018). *Hong Kong inequality report*. Oxfam.

Pavlova, M. (2023). *Green economies, green values: Story for the times*. Kerry J. Kennedy, margarita Pavlova, John Chi-Kin Lee (Eds), *Soft skills and hard values: Meeting education's twenty-first century challenges* (19–36). Routledge. https://doi.org/10.4324/9781003219415-3.

Pavlova, M. (2019, July 14). *Greening of the economy through partnerships: Issues and impacts on skills development*. (S. McGrath, M. Mulder, J. Papier, & R. Suart, Eds.) Handbook of Vocational Education and Training, 195–218. https://doi.org/10.1007/978-3-319-94532-3_116.

Pavlova, M. (2018). Fostering inclusive, sustainable economic growth and "green" skills development in learning cities through partnerships. *International Review of Education, 64*(3), 339–354.

Pavlova, M. (2017). Green skills as the agenda for the competence movement in vocational and professional education. In M. Mulder (Ed.), *Competence-based vocational and professional education. Bridging the worlds of work and education* (pp. 931–951). Springer International Publishing Switzerland.

Payne, G., & Payne, J. (2004). Participant observation. In *Key concepts in social research* (pp. 158–161). SAGE Publications, Ltd.. https://doi.org/10.4135/9781849209397.n35

Pearce, D., & Barbier, E. B. (2000). *Blueprint 6: For a sustainable economy* (1st ed.). Routledge. https://doi.org/10.4324/9781849774239

Prescott, J. A. (1971). Hong Kong: The form and significance of a high-density urban development. In D. J. Dwyer (Ed.), *Asian Urbanization A Hong Kong Casebook* (pp. 11–20). Hong Kong University Press.

Pulichino, J. P. (2007). *Usage and value of Kirkpatrick's four levels of training evaluation*. Thesis (Ed.D.). Pepperdine University. Retrieved from https://www.proquest.com/docview/304707180/8606295BDDF048FAPQ/1?accountid=29075&sourcetype=Dissertations%20&%20Theses

Ranstad. (2023, July 26). *HR trends*. Retrieved December 19, 2023, from New study reveals energy and environment sector tops attractiveness in Hong Kong.: https://www.randstad.com.hk/hr-trends/employer-brand/energy-environment-sector-tops-attractiveness-hong-kong/

Research Office–LegCo. (2021, March 5). *Dongjiang water supply*. Retrieved from Research Publications: https://www.legco.gov.hk/research-publications/english/2021issh19-dongjiang-water-supply-20210305-e.pdf

Saral, A. T., & Pavlova, M. (2025). 17: Green skills for refugees in Hong Kong. In *A modern guide to refugee education*. Edward Elgar Publishing. Retrieved Mar 17, 2025, from. https://doi.org/10.4337/9781803928401.00026

Shi, X., Takanori, M., Takashi, M., Xiaoyu, G., & Ang, H. (2020). Toward sustainable development: Decoupling the high ecological footprint from human society development: A case study of Hong Kong. *Sustainability, 12*(10). https://doi.org/10.3390/su12104177

Siegfried, K. (2022, November 15). *Young Rohingya refugees are helping to turn world's largest camp green*. Retrieved from UNHCR: https://www.unhcr.org/news/stories/young-rohingya-refugees-are-helping-turn-worlds-largest-camp-green?_gl=1*1lb7hgh*_rup_ga*MTIzMTUxNzYwNy4xNzM3MDE3MzA2*_rup_ga_EVDQTJ4LMY*MTczNzAxNzMwNi4xLjEuMTczNzAxODE2NS44LjAuMA..

Stone, G., Barnes, J. H., & Montgomery, C. (1995). Ecoscale: A scale for the measurement of environmentally responsible consumers. *Psychology & Marketing, 12*, 595–612. https://doi.org/10.1002/mar.4220120704

Strietska-Ilina, O., Hoffman, C., Duran Haro, M., & Jeon, S. (2011). Skills for green jobs: A global review. *International Labour Organization*. https://www.ilo.org/wcmsp5/groups/public/%2D%2D-dgreports/%2D%2D-dcomm/%2D%2D-publ/documents/publication/wcms_159585.pdf.

The World Bank. (2009). *EcoCities: Ecological cities as economic cities*. Retrieved from World Bank: http://siteresources.worldbank.org/INTEASTASIAPACIFIC/Resources/226262-1246459314652/Eco2Cities_FullReport_ConfEdition6-26-09_sm.pdf

Thomas, J. (2000). *Ethnocide: A cultural narrative of refugee Detentin in Hong Kong*. Ashgate Publishing Ltd.

Tsang, E., Lee, J., Yip, S., & Gough, A. (2020). The green school award in Hong Kong: Development and impact in the school sector. In A. Gough, J. Lee, & E. Tsang (Eds.), *Green schools globally. International explorations in outdoor and environmental education* (pp. 189–201). Springer. Retrieved from. https://doi.org/10.1007/978-3-030-46820-0_11

T-Park. (2021). *Energy, transformation, community*. Retrieved from https://www.tpark.hk/en/

Trier, J. (1994). *Organizations that help the world: United Nations high commissioner for refugees*. Exley Publications.

UN DESA. (2018, May 16). *68% of the world population projected to live in urban areas by 2050, says UN*. Retrieved from UN DESA–News: https://www.un.org/development/desa/en/news/population/2018-revision-of-world-urbanization-prospects.html

UN-Habitat. (2020). *World cities report 2020: The value of sustainable urbanization*. UN-Habitat.

UN-Habitat. (2017). *New Urban Agenda. Habitat III Secretariat*. UN-Habitat. Retrieved from http://habitat3.org/wp-content/uploads/NUA-English-With-Index-1.pdf

UN Habitat. (2015). *The state ofAsian and Pacific cities 2015*. Urban transformations Shifting from quantity to quality. United Nations ESCAP and UN Habitat. Retrieved from https://new.unhabitat.org/sites/default/files/download-manager-files/TheStateOfAsianAndPacificCities2015.pdf.

UNHCR. (2019). Global framework for refugee education. UNHCR. Retrieved from https://www.unhcr.org/events/conferences/5dd50ce47/global-refugee-forum-pledging-guidance-global-framework-refugee-education.html

UNHCR (2024). Global Trends Forced Displacement in 2023. Retrieved from: https://www.unhcr.org/media/global-trends-report-2023?gad_source=1&gad_campaignid=22932962343&gbraid=0AAAAArCUaL7hu8Ng03YaAnsjN816qCq44&gclid=Cj0KCQjwgKjHBhChARIsAPJR3xe8F9idJ4LD9KR1k6FbtqZbxhGvDpj_bsPB_GpXLcjo4wJijDQkRYwaAvT4EALw

United Nations Environment Programme. (2011). *Investing in energy and resource efficiency*. Retrieved from Cities: Towards a Green Economy: http://wedocs.unep.org/bitstream/handle/20.500.11822/7979/GER_12_Cities.pdf?sequence=3&isAllowed=y

Vecchio, F. (2016). *Asylum seeking and the global city*. Routledge.

Walker, R. M., & Hills, P. (2014, May). Changing dimensions of Trust in Government: An exploration in environmental policy in Hong Kong. *Public Administration and Development, 34*(2), 123–136. https://doi.org/10.1002/pad.1677

Wanner, T. (2014). The new 'Passive Revolution' of the green economy and growth discourse: Maintaining the 'Sustainable development' of neoliberal capitalism. *New Political Economy, 20*(1), 21–41. https://doi.org/10.1080/13563467.2013.866081

WEEE-Park. (2021). *WEEE treatment and recycling facility*. Retrieved from Environmental Protection Department: https://weee.gov.hk/en/background/weee-park-weee-treatment-and-recycling-facility/

While, A., Jonas, A. E. G., & Gibbs, D. (2004), The environment and the entrepreneurial city: searching for the urban 'sustainability fix' in Manchester and Leeds. *International Journal of Urbanand Regional Research, 28*, 549–569. https://doi.org/10.1111/j.0309-1317.2004.00535.x
Wong, W. C., Cheung, S., Miu, H., Chen, J., Loper, K. A., & Holroyd, E. (2017). Mental health of African asylum-seekers and refugees in Hong Kong: Using the social determinants of health framework. *BMC Public Health, 17*(153). https://doi.org/10.1186/s12889-016-3953-5
Yau, Y.-H., Rudolph, V., Lo, C.-M. C., & Wu, K.-C. (2018). Restaurant oil and grease management in Hong Kong. *Environmental Science and Pollution Research, 28*, 40,735–40,745. https://doi.org/10.1007/s11356-018-2474-4

Chapter 2
Opportunity Structures and Agency for Refugees: Importance of Resources

Abstract Although migration is an individual decision, it is a process involving crossing borders and involves state-led institutions. In the case of Hong Kong, Kennedy et al. (*Asian Journal of Social Science, 47*, 534–565, 2019) identify two types of institutions in Hong Kong: formal and informal. Formal institutions include legal and regulatory frameworks that govern political processes, public services, private organizations, and markets. Informal institutions include the 'unofficial' structures that govern relationships within organizations, such as bureaucracies, firms, or industries, as well as everyday cultural practices, value systems, and observed norms of behavior in households or among social groups or communities (Alsop et al., *Empowerment in practice from analysis to implementation*, 2006). Refugees and asylum seekers interact with these institutions on a day-to-day basis. The dynamics governing this relationship influence this group's well-being, behavior, and adjustment to the requirements of sustainable urban development. Therefore, this chapter reviews how formal and informal institutions prevalent in refugees' lives influence their well-being to generate a response that addresses their needs.

The first part of this chapter reviews well-being by focusing on resources and capability approaches. The second section reviews the formal and informal institutions influencing refugee resource allocation. By focusing on opportunity structures and agency, the chapter argues that this group needs to be empowered through education and employment.

Keywords Wellbeing · Resources and capability for refugees · Formal and informal institutions in RAS' lives

A. T. Saral, M. Pavlova, *Education for Refugees and Sustainable Green Urban Development*, Education in the Asia-Pacific Region: Issues, Concerns and Prospects 78, https://doi.org/10.1007/978-981-95-1680-3_2

2.1 Section 1: Wellbeing, Resources & Capability in Refugees' Lives

2.1.1 Wellbeing

Wellbeing is a concept that refers to people's capacity to live healthy, creative, and fulfilling lives. There are two dominant interpretations of what constitutes a good life and what makes life fulfilled: eudemonic and hedonic (Western & Tomaszewski, 2016).

The eudemonic view stresses the meaning of life, self-realization, and personal growth to attain optimal well-being. From this view, quality of life is determined by material resources (such as income, food, and housing) and attributes (such as education, health, political voice, social networks, and connections) (Kuan et al., 2010). These are essential elements for ensuring similar levels of well-being across all people; however, human beings also have psychological needs; they need to feel they belong and need development. For this reason, this view falls short of a comprehensive understanding of well-being, as the experiential side of human beings is not factored in (Veenhoven, 2008).

From the hedonic point of view, life involves pleasure and pain, finding happiness by attaining goals or valued outcomes in varied realms from an individual's perspective (Veenhoven, 2008). One of the theories about this view suggests that people's appraisal is culturally constructed. From this perspective, the shared notions of social well-being can be observed. However, when people are asked to evaluate their lives, they tend to use their current mood as the prime source of information (Veenhoven, 2008). Therefore, the overall judgment correlates to feeling in a given instant, more than general contentment. Another theory about the subjective interpretation of well-being suggests that the perception is based on social comparison and considers the gap between life as it is and how life should be. However, social comparison tells us what is possible in life, not necessarily what is enjoyable or desirable. For this reason, Western and Tomaszewski (2016) argue that components of subjective assessments of people's feelings can help assess objective well-being.

The relationship between objective and subjective well-being is meaningful, but the essence of the relationship is hard to characterize because many factors influence well-being, such as financial and personal security, meaningful and rewarding work, good health, etc. The two viewpoints, hedonic and eudemonic, overlap and complement each other. Therefore, a balanced approach to well-being is conceived as a multidimensional phenomenon that includes individual and objective views (Western & Tomaszewski, 2016; Ryan & Deci, 2001). In terms of this idea, Sen (1999) suggests there are many aspects of people's lives that impact their well-being, such as resources. Sen's capability approach suggests that individuals differ significantly in their abilities to use or make sense of their resources. Therefore, an evaluation of well-being would be incomplete without a consideration of what particular people can do with their resources because even if the distribution of resources is fair, abilities are not. In other words, the lives of people with the same vision and the same resources can still result in completely different lives. Sen (1999) argues that the relationship between people and resources should be examined if we are to enable people to contribute to society equally and understand how this influences their well-being. The essence of this argument is not about how equally capable people are but about how equality is achieved. This aspect should also be incorporated into an evaluation of well-being.

2.1.2 Resources

The concept of resources, when it comes to refugee-related studies, is understood in terms of an individual's needs, personal goals, and the demands they encounter. The difference between resources and goals is that resources, such as money, energy, and education, can be used for different goals by different individuals (for example, a person whose goal is to become a professional diver or another person whose goal is to become an astronaut, use their resources (e.g., education) differently. The desired satisfaction of needs and goals places demands on an individual's resources, or they can result from unforeseen or external events. Ryan et al. (2008) have studied the resources refugees possess. Personal resources are related to physical or psychological needs; material resources are related to money, property, and personal possessions, including employment; social resources relate to one's emotional, informational, and tangible support in personal relationships; and cultural resources are defined by one's skills, knowledge and beliefs within a cultural setting (Ryan et al., 2008).

A meaningful life is about determining and planning to achieve those goals. For instance, a refugee may spend hours learning the local language to increase their chances of securing employment—but this remains dependent on how local authorities act. The notion of capability, in this regard, focuses on the agency dimension, that is, the capability to participate in society and the ability to choose (McGrath & Powell, 2016) because resources alone do not have a value unless they are connected to a particular human functioning. Entering a new socio-economic setting, as refugees do, inevitably causes loss or a devaluation of some cultural resources, education, occupational skills, and experience because these may not be relevant in the new place. Furthermore, refugees' adaptation depends on their ability to replace or substitute their resources to regain lost or gain new ones. Hobfoll (2001) describes this coping mechanism as an investment in resources. In a refugee's situation, the loss of social support from family and friends in the home environment can be replaced by organizations, new friends, and community in the host environment.

On the other hand, substitution involves compensating/exchanging a resource with an equal or less valuable resource from another domain. For instance, the often-observed lack of self-esteem in the refugee community due to unemployment, particularly amongst women, can be substituted by participating in training and helping them realize what they can do if given a chance. Such opportunities create an environment where they can socialize, support one another, and establish new skills relevant to the host environment. Capacity-building interventions are argued to be particularly effective for those with lower levels of well-being (van Agteren et al., 2018). Concordant with the positive impact of skills training on well-being, a study in Iran also found that life skills training improved female students' subjective well-being (Sadr-Mohammadi et al., 2014).

2.1.3 Capability

Learning and development are fundamental factors that assist human beings in reaching their potential, being a part of change, and progressing as individuals, cultures, and societies. Sen (1999) uses the term 'agent' as a person who acts and brings about change and whose achievements are evaluated based on their valuation and reasoning. With rapid global urbanization, international studies highlight the importance of understanding the changing urban environment, how to treat urban systems, and their impact on human well-being (Bai et al., 2012; McGranahan et al., 2005). Urban societies need to realize this aspect by linking well-being with the environment. Facilitating this understanding can be established through school programs, the works of non-governmental organizations and government agencies, and the communication of new methods and technologies to the public (Kellert, 2012). However, none of this can be achieved and made sustainable unless people develop environmental values, ethical responsibility, knowledge, love, and motivation to experience the environment. This is what generic green skills are about: making people understand sustainability and the challenges and opportunities it brings, not simply in economic terms but also societal terms, because, at their core, these skills involve developing interpersonal and intrapersonal skills. Achieving an environmentally sustainable urban development necessitates the development of ecological and environmental awareness, considering energy use and natural resources, and generating and managing waste, all of which generic green skills can foster. On the other hand, social enterprises are considered a platform where all of it can be achieved with a more significant impact on communities. Therefore, enabling migrants and RAS to contribute to sustainable development through green skills could benefit Hong Kong by promoting lifelong learning, achieving sustainable urban areas, and empowering the RAS community. Through training for social entrepreneurship competencies, RAS can showcase their creativity, ability, and resources in an environmentally and socially conscious way and benefit society.

In trying to access resources, people, including RAS, interact with institutions and communities. Kennedy et al. (2019) identify two types of institutions in Hong Kong: formal and informal. Formal institutions include legal and regulatory frameworks that govern political processes, public services, private organizations, and markets. Informal institutions include the 'unofficial' structures that govern relationships within organizations, such as bureaucracies, firms, or industries, as well as everyday cultural practices, value systems, and observed norms of behavior in households or among social groups or communities (Alsop et al., 2006). In the case of the RAS community, however, formal and informal institutions may limit asylum-seeking and refugee groups their ability to exert resources and skills. Sen (1999) also acknowledges that social, economic, and political opportunities can constrain effective freedom: freedom to choose from available options that a person values. For example, women worldwide lack opportunities and support to attend (technical) education, work, and politics. This type of impeded opportunity to cultivate cognitive and imaginative faculties negatively influences well-being (Nussbaum, 2000). In addition, this situation leaves people with little room to make choices and practice for a reasoned agency (Sen, 1999).

In his capability approach, Sen (1999) focuses on quality of life and analyzes it in terms of "functionings and capabilities." Functioning is a state of being and may include, for example, being well-nourished and having shelter. On the other hand, capability refers to valuable functionings a person has effective access to. In other words, a person's capability refers to their freedom to choose from available options—functioning combinations—and to attach a value to them. In this approach, resources are inputs individuals value to choose based on their reasoning. The utility is both an outcome and a functioning. This is because what people choose to do impacts their sense of well-being. For this reason, it has a valuable function (Wells, 2012). For example, an agent can value a specific functioning of a resource and choose to utilize it in a particular way. Let us say an agent can access a piece of leftover fabric (resource). They are either able to sew it or not (capability). Let us assume they can sew or are willing to learn how to sew it. In this example, the person can choose to sew leftover fabric for different functionings, such as making a wallet or a toy for their child. Therefore, the functioning of a resource is decided by the agent's reasoning-valuation. They become happy if they enjoy the activity (either sewing or learning to sew). This happiness adds to their subjective well-being from a hedonic point of view. Table 2.1 is created to facilitate an easier understanding of terms related to this approach.

From a eudemonic approach, on the other hand, certain aspects impact well-being. These are economic standards, equality, the natural environment, and human health and knowledge. Sen argues that development should go beyond maximizing income, accumulating wealth, and gross national product (Sen, 1999). Development should increase one's quality of life, enhance freedom, reduce inequality, and guarantee inclusion regardless of age, sex, disability, race, ethnicity, origin, religion, or economic status. People's ability to choose which resources they would like to use and the contribution they would like to make to society is the central notion of sustainable and human development (McGrath & Powell, 2016). In line with this

Table 2.1 Terms related to the capability approach

Resource	Functioning	Capability	Utility
Leftover fabric	Making a wallet or toy	To sew—To learn how to sew	Pleasure of doing/learning, sense of happiness

reasoning, SDG 10 states that development is not sustainable if people are excluded from opportunities, services, and chances for a better life.

The ultimate goal for people is to find happiness and/or self-actualization (Alsop, Bertelsen, & Holland, 2006). So, based on the above example about waste fabric, the provision of or access to leftover fabric (resource) is necessary to discuss capability in the first place. Doing something that the person values with it, e.g., taking a sewing course or making a toy/wallet with it (functioning and capability), follows because that means they have choice. If, in the end, they feel happy and/or experience personal growth due to their choice, then a certain level of freedom is achieved. So, if a person can choose, it increases their freedom. Therefore, allowing groups like refugees to access training programs can allow them to choose to attend or not attend and expand their sense of freedom. However, when assessing capability, one should consider the number of options available and the freedom to choose a high-quality life. For this, the valuation of the available functionings should be improved. To address this in a way that refugees may value, we also need to understand their main issues and challenges. Below, we will touch on the issues refugees and asylum seekers were reported to face in Hong Kong based on the literature review. These experiences are reported based on different methods and in various terms, but we will attempt to categorize them based on the concept of resources framed by Sen's approach. It is essential to note that they are intricate and hard to categorize because either the people themselves or their effects overlap. For example, a material problem may be the reason behind a personal (e.g., psychological) problem.

In terms of personal resources that cover areas related to physical and psychological well-being, one study that looked into refugees' vanishing identity in Hong Kong found that the long waiting period for claim assessments causes refugees to fail to envision their future. Furthermore, this unforeseen future is reported to be psychologically disturbing. Moreover, the limited support for government food coupons was also considered to have left the refugee community destitute (Lau & Gheorghiu, 2018). In another study that investigated the mental well-being of African asylum seekers in Hong Kong, it was found that a large proportion of them expressed depressive symptoms (Wong et al., 2017). These symptoms were argued to have caused some of them to lack a sense of security in Hong Kong (Ng et al., 2019).

Material resources are objects and properties a person possesses, such as money (including employment), cars, houses, etc. In this regard, Kennedy et al. (2016) study found that the two main issues for the refugee community in Hong Kong are financial problems due to the inability to work and the other is housing (i.e., accommodation). However, as Vecchio (2016) highlighted, many refugees take up illegal work because otherwise, sustaining life in Hong Kong is challenging. Some become involved in recycling or trade activities. In terms of recycling jobs that serve to green the city, it is mainly men who handle bulky car parts or electrical home/office appliances thanks to their stronger physiques. In terms of trading, they use their ethnic and personal networks in their home country or elsewhere to trade products such as mobile phones, clothes, watches, computers, and other small electronic (counterfeit) goods. Some refugees were also reported to instruct newly arrived African traders about trading (Mathews, 2018 cited in Choy & Shi, 2021).

Regarding social resources that relate to one's social network, information, and emotional support mechanisms, refugees stated they have experienced unfriendliness or even hostility on the street daily. They experience difficulties seeking help and services in private and public areas, which can ultimately be discrimination (Ng et al., 2019). Most commentators writing on the matter supported this aspect of discrimination and exclusion from society. Shum (2014) also considers refugees' experiences in the pre-flight, flight, and post-arrival periods and explains the roles of previous migrants or mediators and smugglers in their arrival experience in Hong Kong. He notes that those who arrived in Hong Kong via smugglers demonstrated a qualitatively different experience from those who arrived via legitimate air travel (Shum, 2014), thereby indicating that refugees' experience before arriving in Hong Kong can influence their use of resources and their experiences in the society once they have arrived.

Cultural resources relate to one's knowledge and skills. For example, Kennedy et al. (2016) note that language barriers can negatively influence refugees' lives. Their inability to understand the language leads to misunderstanding or poor communication with various institutions, such as the Immigration Department (ImmD) or Education Bureau. As a result, they may feel powerless and vulnerable as they attempt to negotiate in a language they do not understand. On the other hand, Matthews et al.'s (2014) study demonstrates differences in refugees' perceptions. In his study, asylum seekers were reported to be proud of their ability to outwit the state by working illegally—which, to them, is legitimate (Matthews et al., 2014). Some asylum seekers of African origin who married a local Chinese wife were reported to be able to penetrate the labor market and acquire new social networks and cultural capital. However, regarding access to training and education, the state mechanism follows a restrictive approach and does not provide refugees with skills training. This situation contravenes the idea that training and education improve people's agentic capabilities and transform their world (Powell, 2012 in McGrath & Powell, 2016).

2.2 Section 2: Formal and Informal Institutions

2.2.1 Policies, Regulations, and Social Interactions Concerning Refugees and Asylum Seekers in Hong Kong

As Hong Kong is not a signatory of the United Nations Convention relating to the Status of Refugees, illegal immigrants who claim that they might face torture or inhuman treatment in their home countries are not treated as "asylum seekers" or "refugees." The government set up a Unified Screening Mechanism (USM) in March 2014 to screen these nonrefoulement ("NR") claims on all applicable grounds. As of October 2023, around 14,700 non-refoulement claimants remained in Hong Kong for different reasons (Legislative Council, 2023). The claimants are mainly from Southeast Asia, such as Indonesia, India, Bangladesh, and Vietnam (HK Immigration Department, 2024). Out of 30,000+ applications in the last 15 years, only 349 were substantiated (HK Immigration Department, 2024).

Hong Kong's approach to claimant-related policies has been criticized locally and internationally for being strict and limiting (see Kennedy et al., 2016; Kennedy et al., 2019; Vecchio, 2016; Refugee Concern Network, 2021; Justice Centre Hong Kong, 2020). As a result, it is crucial to understand the push-and-pull factors behind asylum seekers' choice to migrate to Hong Kong to address their needs better. *The push factors* are political and religious threats, persecution, and economic problems in the country of origin. *The pull factors* are freedom, safety, and security about places, jobs, and study opportunities in Hong Kong (Kennedy et al., 2019).

Hong Kong also pulls asylum seekers for religious reasons. Often, when converting from one religion to another, people face community persecution. However, Hong Kong pulls asylum seekers due to its visa-free policy (for up to 180 days) and its granting of freedom to practice religion.

Economic reasons, in terms of push factors, include high unemployment rates, famine, disease, and natural disasters. When the financial situation in a home country no longer provides the opportunities people expect, asylum seekers go to economically developed countries or big cities such as Hong Kong to seek jobs and a better life (Kennedy et al., 2016). Therefore, Hong Kong's economically advanced status in the region and job opportunities are often considered pull factors for those from India, Nepal, and Bangladesh seeking employment. Lawmaker Simon Lee Hoey said, "*When we look at the countries of origin of those people, most of them aren't from places under active armed conflict, but mostly from economically less developed areas within the region ... most of them don't truly fulfill the criteria as a refugee*" (Ma & Hamer, 2024). However, human rights lawyer Mark Daly argued that concerns of abuse should not discredit the availability of legal channels or the legitimacy of asylum seekers' claims as a whole, as the system was set up to sift through applications, including bogus ones (Ma & Hamer, 2024).

Political instability and terrorism in other countries are other push factors for RAS. As of September 2024, determinations have been made in 30,399 torture/non-refoulement claims, among which 349 were substantiated, indicating a very low level of approval rate. A breakdown by nationality of the substantiated claimants shows that claimants from Pakistan and Yemen are number one in substantiated claims, with 43 supported claims in Hong Kong (HK Immigration Department, 2024). For example, Yemen has experienced civil war since 2015 between the government forces and the Houthis, a minority Shiite group (Riedel, 2017). Although there is no data concerning granting substantiated claims in Hong Kong, other countries that RAS come from, such as Somalia, Egypt, and the Democratic Republic of Congo, are also politically unstable, economically corrupt, and hit by famine. So, Hong Kong's comparatively safer and stable environment can be considered a pull factor. In some cases, as explored by Kennedy et al. (2016) and Shum (2014), people were cheated by traffickers with the promise of finding a job in Hong Kong.

However, in public perception and government policies alike, this group of people is often labeled "illegal immigrants" or "economic immigrants." In international parlance, these terms define different groups of people. An illegal immigrant refers to migrants in irregular situations[1] or undocumented migrants. An economic migrant refers to anyone who has moved across an international border or moved within a state solely for perceived economic opportunities. On the other hand, an asylum seeker is an individual applying for international protection and seeking to be recognized as a refugee by local authorities in another country. Understanding this delineation is important for several reasons:

- These definitions are used in international law to create a common denominator across states, contributing to effective collaboration across relevant parties.
- The word ‘illegal’ has a criminal connotation, especially in the case of migrants in an irregular situation; it also undermines the dignity of migrants.
- The categorization of migrants being motivated by a single reason, such as economic, is inaccurate because migration involves overlapping and complex drivers.
- The use of “economic migrant” instead of “refugee” and “asylum seeker” is contrary to the prescriptions of international law. It overrides the universality of human rights because refugees are entitled to specific protection under refugee law, whereas states have no obligation towards migrants who seek better economic opportunities abroad (IOM, 2019).

The blurry part in Hong Kong’s law occurs when a migrant is in an irregular situation or when an overstayer lodges an application under the Convention Against Torture (CAT) for asylum. Therefore, in Hong Kong, the status of an asylum seeker, who is treated the same as other arrivals in Hong Kong, overlaps with that of migrants in an irregular situation, and they are locally referred to as illegal immigrants (Vecchio, 2016). Moreover, a brief consideration of the unified screening mechanism (USM) demonstrates that to apply for asylum, one must overstay their visa first. This means the applicants are already classified as *immigration law offenders* or *criminals* in Hong Kong’s system. In addition, governmental sources often address RAS, in local terms, as claimants and divide them into two groups: overstayers and illegal immigrants.[2]

Most of the overstayed people or illegal immigrants traveled to Mainland China from their country of origin as a first point of entry. They traveled by land or flight and arrived at the coastal area of Guangdong or the land boundary of Shenzhen before smuggling into Hong Kong by sea or over land. Among claimants who overstayed or were refused permission to land upon arrival, almost half were visitors from visa-free countries, and about 70% were from India (Security Bureau, 2018). The government started implementing specific measures related to non-claimant’s pre-arrival to prevent people from abusing the system and being smuggled into the

[1] Irregular migration—Movement of persons outside the laws, regulations, or international agreements governing entry or exit from the state of origin, transit, or destination (IOM, 2019).

[2] Current statistics show that 59% of claimants are overstayers, 33% are illegal immigrants, and 8% are others (including persons refused entry and persons born in Hong Kong with the right of abode in Hong Kong not yet established) (HK Immigration Department, 2024).

city. Examples are the pre-arrival registration form for incoming Indians, the government's stricter penalties for overstayers with jobs, and employers who hire illegal workers (Leung & Lok-kei, 2019).

Although Hong Kong wants to avoid attracting refugees, there is no benefit in blank rejection resulting in a negative reputation internationally. The literature review indicated that accepting refugees in Hong Kong is not based on humanitarian concerns alone. Political grounds play an essential role in this process. These concerns can set the foundation for opportunities for incoming groups of people. Therefore, they can be highly influential regarding their settlement and inclusion in the host societies. However, it is acknowledged that *policies in place* (output) and the *implementation of these policies* (outcome) are not always linear (Norman, 2018). For example, while repressive policies make it very difficult for migrants and refugees to obtain residency and membership in society by any means, in liberal regimes, which are often democratic states, refugees' and migrants' residency is allocated after a certain period, and they are permitted to participate in the national economic system.

From an international political perspective, some refugees or asylum seekers may be more 'desirable' than others for historical, political, economic, or diplomatic reasons. A recent example is the European Union, where Ukrainians were granted immediate refuge. However, the same host countries have neglected Syrian refugees for a long time (-though it started in 2011 and Turkiye has hosted millions, the EU acknowledged it as a crisis only in 2015). Turning to Hong Kong, 40,000 Vietnamese refugees were detained there as part of the British Government's foreign policy at the time—when Hong Kong was an appendage (Thomas, 2000).

Looking through the local political lens, Hong Kong is a special administrative region of China. It retains its power to arrange an immigration policy under the "one country/two systems" framework. This is somewhat similar to what can be observed in Europe. The Union has a policy towards immigration and integration and can provide incentives and support to its member states; however, states can develop their own policies. So, Hong Kong's approach to attracting global talents and maintaining regional mobility of labor is the opposite. For example, Hong Kong welcomes and actively engages in worldwide talent hunts for high-skilled professionals to keep up with economic and technological advancements, such as attracting people from developed countries like the USA or the UK. On the other hand, foreign workers from Southeast Asia are strictly regulated, and they confront many barriers, and their stay, due to policies, remains temporary (Vecchio, 2016).

Focusing on refugee policies in Morocco, Turkiye, and Egypt, Norman (2017, 2018) investigated transit states' engagement with non-citizens, a category under which refugees fall. She looked at the relationship between the host states and refugees concerning inclusion in residency and membership, as well as access to health care, education, housing, and employment. In her findings, she introduces the policy of indifference as a way of engagement. For this type of engagement, host states acknowledge the existence of refugees and migrants, but they refrain from directly engaging with them. Instead, they rely on and are willing to collaborate with (international) organizations and NGOs to directly engage with the group concerned as it may benefit the states diplomatically or economically. Under a strategy of indifference, refugees' or migrants' residency is either permitted for a short time or authorities frequently check it, and employment in the formal economy is not allowed. Nonetheless, work in an informal economy will remain pervasive and is not highly policed (see Norman, 2018, 2019). This engagement alleviates the host state's responsibility while ensuring that migrants and refugees are not excluded from essential services, such as access to health care, education, and employment (Norman, 2018, 2019). Hong Kong's engagement with RAS regarding residency, housing, social interaction, access to basic needs (health services and education), and employment influence RAS's well-being are reviewed below.

Regarding residency: While a skilled and financially well-off foreigner can apply for permanent residency in Hong Kong after 7 years of continuous residence, foreign domestic helpers such as those from Indonesia are excluded from such opportunities (Vecchio, 2016). There is one case of a domestic helper who had worked for 25 years, but the permanent resident application was overturned because it was deemed unconstitutional (Lau, 2012). The government consistently emphasizes that regardless of the outcome of a person's claim, their immigration status as illegal immigrants/overstayers will not change, and they are not permitted to stay in Hong Kong[3] (Security Bureau, 2019). The purpose is to separate them from the public, exclude them from the workforce, and monitor them closely. These decisions and implementations highlight immigrant class divisions and support Norman's argument about the indifference strategy.

Concerning access to employment, Article 37ZX of the immigration ordinance states that claimants of substantiated claims may apply for work, and the decision is at the discretion of the Director[4] and will be based on an exceptional circumstance that can be justified (Cap. 115 Immigration Ordinance, 2021). Although this is considered an improvement by authorities, there are still problems, such as:

It may take years for a case to be substantiated, during which time the person is not allowed to work and may be subject to removal. Therefore, in addition to legal restrictions, logically, if one is subject to removal from a country at any time, one cannot anticipate securing a position. The government has indicated that the time for handling each claim has been reduced from 25 weeks to 10 weeks by the United Screening Mechanism (USM) thereby specifying an improvement in terms of output (HKSARG, 2020). However, following these handlings, many non-refoulement claims challenge the USM decision and apply for a judicial review (JR), revealing a low efficiency (See Image 2.1). Only 1.2% of JR claims over the past 15 years were substantiated (The Legislative Council Commission, 2024).

---

[3] They will not be treated as refugees or asylum seekers. Substantiated claimants can stay until their claimed risk ceases to exist or they should relocate to another accepting country (Security Bureau, 2019).

[4] Director refers to the Director of Immigration, the Deputy Director of Immigration, any assistant director of immigration and any member of the Immigration Service with the rank of senior principal immigration officer (HK Immigration Department, 2021).

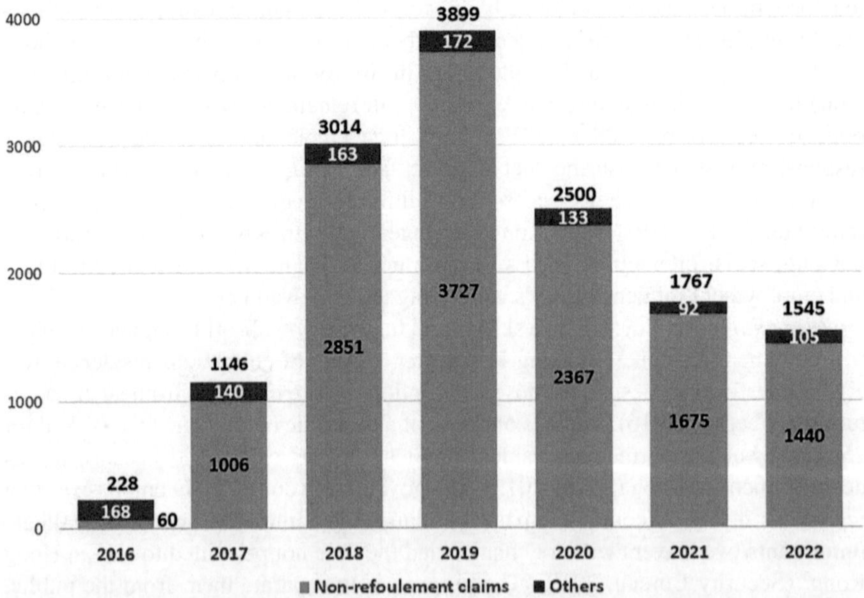

Image 2.1 Number of applications for leave for JR. Source: The Legislative Council Commission, 2024

Government statistics (Secretary for Security, 2021) show that the Immigration Department (Immd) received 368 applications for work permits, among which substantiated claimants lodged 187 and 181 were lodged by persons whose refugee status had been recognized by UNHCR. Among the 368 applications received, 295 were approved, 10 were rejected, four were pending further information from the applicants, and 59 were withdrawn or where no further action could be taken (Secretary for Security, 2021). This indicates an approximately 80% approval rate among those who apply. However, according to a local NGO working closely with refugees and asylum-seeking communities, the approval rate is lower (Branches of Hope, 2024). Therefore, we see a discrepancy and difference in numbers and perception between the government and NGOs, which confirms Kennedy et al.'s (2019) finding of the discourse. There may be additional reasons that prevent RAS from applying. These may be related to external causes such as the volume of documentation required for a work permit, a lack of employers willing to go through this process, or internal reasons such as lack of self-esteem, knowledge, and skills, which in turn highlight the necessity of empowerment for this community.

Moreover, according to SCMP (2012) news, five asylum seekers who were fighting to gain work were reported to be depressed and suffering from other mental diseases (SCMP, 2012). Therefore, under these circumstances, even if the court accepts a claim as an exceptional case, would an employer hire someone with mental disease?

Although it can be argued that a case-by-case basis may progress a humanitarian approach and, at the same time, prevent abusers from taking advantage of the situation, it can also be due to a lack of standards.

On the policy level, asylum seekers are strictly forbidden from engaging in the formal economy (Ramsden & Marsh, 2013). Vecchio's (2016) study demonstrates that although refugees are not allowed to work in Hong Kong, some local entrepreneurs do generate income by engaging in the recycling of electrical and electronic waste or by targeting low-end demand for inexpensive clothing and food, and this is where the asylum community often works. So, as opposed to policy, the refugee and asylum-seeking community penetrates the informal economy; it becomes part of the business chain and is then exploited (Vecchio, 2016). This point also supports Giddings et al.'s (2002) argument regarding sustainable development, where legal and non-legal systems cannot be separated, and there are multiple ways for societies and economies to function (Table 2.2).

Table 2.2 indicates that refugees are provided with a maximum of HK$ 3490 per month. The poverty line in HK, according to 2020 statistics, was HK$ 4400 (Census and Statistics Department, 2021). Therefore, as Vecchio (2016) explores in his study, it is almost impossible not to need work, given the poor conditions refugees are expected to live in.

Kennedy et al.'s (2019) example of a systematic fault within the policy and its implementation illustrates this fault well. He mentions that if refugees are caught working illegally, they are incarcerated. In prison, they can work and earn a living of up to HK$600 per month (Kennedy et al., 2019). This situation can be described as a strategy of indifference Norman (2019), whereby refugees are de facto permitted to work under certain circumstances.

A historical example of indifference can be given in terms of housing: many Vietnamese refugees were de facto integrated into society during the 1970s and '80 s and worked to support themselves. Their accommodation was outside the Pillar Point Vietnamese Refugee Centers (PPVRC). It was established by a private company—the Hong Kong Housing Services for Refugees Ltd.—and was managed by Caritas Hong Kong, an international charity organization. Although a local resettlement scheme was introduced in 1986, it was unsuccessful. Only 196 places were taken out of the 250 available (Legislative Council, 2000). The then Governor-in-Council approved the Local Resettlement Scheme on 6 February 1986. Vietnamese refugees who met the following criteria were eligible to apply for settlement in Hong Kong:

(a) they had been in Hong Kong since July 1982;
(b) they were ethnic Chinese;

Table 2.2 Humanitarian assistance available for claimants

Item	Amount (HK$)
Food	$1200 per month
Rent	$1500 per adult and $750 per child per month
Utilities charges such as electricity and gas	$300 per month
Transportation	$200to $420 per month
Other basic necessities	$70 per month, provided in-kind

Source: The Legislative Council Commission, 2022

(c) they were financially independent and
(d) they did not meet resettlement criteria in other resettlement countries.

This scheme has a monthly quota of 20 and an overall quota of 250 (Legislative Council, 2000). Despite being introduced as an integration policy, the criteria (b) and (c), in particular, create a restriction for eligibility because RAS "often encounter difficulties, for example, in finding a job, because of their status and because they do not have equal access to various services and facilities as ordinary Hong Kong citizens do" (Legislative Council, 2000, p.5). This example concerns housing, a problem many refugees still face today.

Regarding health services in Hong Kong, refugee children are entitled to access healthcare services at public hospitals and clinics (which are managed by the Hospital Authority and the Department of Health). The charges for this service may be waived following an application for a medical waiver through medical social workers (at the Hospital Authority) or social workers at Integrated Family Service Centers (IFSC) (at the Social Welfare Department). Unfortunately, practical and administrative barriers often prevent refugee children from enjoying these rights. Some refugee children require regular medical appointments. However, the children and their parents are not given additional transport allowance from International Social Services (ISS) to finance their journeys to and from hospitals and clinics. The outcome of this is that children and parents must often choose between eating and attending essential medical treatment.

Similarly, school children receive free dental care up to the sixth grade. However, children in secondary school have no access to dental care (apart from tooth extractions). Some children are reported to have suffered from severe dental pain that they cannot attend school. Unlike many other families, the families of refugee children do not have the resources to pay for private dental care. Public dental clinics (managed by the Department of Health) only provide emergency dental services (limited to pain relief and tooth extractions only). Limited quotas are given on a first-come-first-served basis; therefore, whether treatment is received at all or in time leaves a question mark concerning the outcome (Save the Children Hong Kong et al., 2017).

Regarding policies related to *access to education*, the government ensures admission to public schools for all eligible students, including non-Chinese speaking children, under which claimants fall. However, the Equal Opportunities Commission submitted a case to the Legislative Council regarding the fact that only one out of 16 Vietnamese secondary school students were given a permit. At the same time, the rest waited for 3 months to gain permission to attend school from the authorities, namely the Immigration Department and Education Bureau (Equal Opportunities Commission, 2016). This is an outcome of a reluctant approach by the authorities.

Regarding *interactions with locals* in Hong Kong, claimants' children can interact with peers thanks to the school environment. However, adults' lack of access to work or training opportunities makes interactions with locals challenging, particularly for women, so they remain confined to their community. This situation can be explained through the divisive and indifferent policy mechanics of the administration in Hong Kong. A study about the discrimination experience based on the residency status of RAS in Hong Kong suggested that approximately 90% of this population indicated they had experienced discrimination at the daily, operational, and policy levels (Ng et al., 2019).

The section above established that refugee groups' participation in society is systematically oppressed and stigmatized in Hong Kong due to existing policies and their implementation, indicating the need to empower this community.

The existing laws and regulations influence the participation and use of agency by individuals to develop different capitals[5] and resources. Esser (2004) argues that immigrants' investment to develop their capital and resources depends on the perceived success rate and the investment cost, choosing between a secure gain and a risky and costly investment. The types of investments Esser (2004) defines are:

- *Investment in resources and human capital* directly related to the receiving country, such as formal education
- *Investment in ethnic group capital* for economic advancement, such as the opening of ethnic markets
- *Political investment* in favor of a re-evaluation of the ethnic group capital already in stock, such as ethnic movement

According to Esser (2004), the expectation for potential success is crucial because it is this expectation that helps overcome risks and indicates a high motive for investment. Investment in specific resources depends on the correlation between expected success probability and cost. While success expectation is related to the availability of information, which increases with the duration of RAS's stay, overcoming the cultural and social distance requires resources. For example, Larsen et al. examined refugee outcomes regarding educational investment and labor market participation. According to their findings, women refugees are more likely to be students (either in formal or other types of education that may enable their access to the labor market) rather than working or being unemployed (Larsen et al., 2018). These authors suggest that women tend to substitute work for education. Alternatively, they may have difficulty securing a job, so they choose education out of necessity rather than choice.

On the other hand, a study from Sweden examined the effect of permanent residency—a formal institution—on the labor market participation among refugees in the short term and found that temporary residents subject to a relatively less inclusive situation have higher incomes and less unemployment. However, at the same time, they are less likely to spend time in education than those with permanent residency (Jutvik & Robinson, 2020). So, the certainty of protection (through temporary or permanent residency) influences refugees' investment in education, representing an interplay between formal and informal institutions.

The above examples demonstrate that investments in terms of the use of resources, strategies, and functions are influenced by economic opportunities in the

[5] By capitals, we mean the skills, knowledge, experience, and education of people (Ungera, Rauchb, Fresec, & Rosenbusch, 2009).

host society or cultural reasoning/values. Investment in ethnic group capital increased with market opportunities and the productivity of the ethnic businesses. As a result, entrepreneurial experience and the availability of ethnic and social capital increase the probability of success. Costs decrease with an increased opportunity for exploiting ethnic solidarities (Esser, 2004). Paradoxically, due to the refugee population's illegal status, lawfully resident ethnic employers exploited their fellow countrymen when they provided labor in Belgium (Grzymala-Kazlowska, 2005). Although it is unlawful for asylum seekers to work in Hong Kong, it is impossible for them not to work, given the minimum support they receive from the government (Mathews, 2018). This indicates that refugees participate in the social or economic realm. However, Vecchio (2016) finds that refugees are exploited by locals or by their countrymen when they run a café or market in Hong Kong. Some were reported to be self-employed entrepreneurs who serve tourists or tourist traders from developed or emerging countries. So, often, the choice is between being self-employed entrepreneurs or working long hours in low-end jobs, which increases their vulnerability because it makes it difficult for them to break these ties and gain legal status and secure socio-economic prosperity (Vecchio, 2016). In terms of gendered experiences, Thomas investigated Vietnamese detainees in camps in 2000. His findings indicated that the role of the mother—the one who provides nourishment—was weakened due to the refugee experience in camps in Hong Kong (Thomas, 2000). Although there are no longer refugee camps in Hong Kong, a paper that focuses on migrant domestic workers (MDWs) highlighted the struggles that some MDWs face; they are victims of human trafficking and applied for asylum in the city (Anderson & Li, 2018) and are in dire need of empowerment and recognition.

Regarding political investment, Dahl's study looks at the relationship between resources and investment at the political level—as this is the level where one exerts influence over decision-making. According to this study, the more resources one has (income, education, social standing, occupational standing, and so on), the more likely one is to participate in local decision-making (Dahl, 1961). However, Dahl also notes that individual decisions may vary concerning the use of resources for political reasons due to personal situations (such as goals, values, identification, information, etc.) or objective situations (poor, uneducated, unskilled labor, living in slum or lower-class neighborhoods, etc.). Thus, the study indicates that although resources are essential in decision-making, agents do not necessarily mobilize them in the decision-making process. Hence, resources only represent power potential, not necessarily the degree of power itself (for more details, see Power Concept in Appendix 2, Alsop et al., 2006).

2.3 Summary

The literature review indicated that public policies and practices remove asylum seekers from public sight by confining them to desolate socio-economic and political spaces (Vecchio, 2016). Thus, although some inputs by the system, e.g., faster handling or eligibility to apply for a work permit after a case is accepted, have improved, the outcome is limited.

Situations related to power (the ability or capacity to do something), resources (things that have utility and add value to life), and the motivation to invest in resources are often associated with asset-based approaches and are expected to bring economic advancement. However, the very reason people apply for asylum is not generally only for financial reasons, which indicates the importance of other resources in their lives, such as social or religious resources. Nonetheless, the lack of economic opportunities in the destination country can motivate their decision to participate or not participate in the financial or social realm.

The literature review also indicated that institutional structures (such as governmental regulations) and individual ability, or capacity to act, influence each other. Therefore, potential financial outcome is not the only way to empower refugees and motivate them to become change agents. By "agents of change," we refer to the recognition that people have an agentic capacity and can use it positively and effectively and take advantage of the options available. If they use their agentic capacity, they can help the development of themselves and others (Szekely & Mason, 2019) because '*capable people have confidence in their ability to take effective and appropriate action, explain what they are about, live and work effectively with others and continue to learn from their experiences, as individuals and in association with others, in a diverse and changing society*' (Stephenson, 1998).

The situation in Hong Kong, in particular, indicates that the interplay between formal and informal institutions can encourage refugees to find ways to recognize their resources and apply their agency to make use of them, such as working in shops run by people of their ethnicity (in a way, this is empowerment because they cannot survive otherwise so they use their cultural resources to locate that job). However, this very situation limits their freedom and marginalizes them further since it makes them more vulnerable to exploitation. Therefore, RAS empowerment is highly necessary.

The next chapter focuses on empowering refugees and the need for capacity-building through green skills development.

References

Alsop, R., Bertelsen, M. F., & Holland, J. (2006). *Empowerment in practice from analysis to implementation*. The World Bank. Retrieved from https://elibrary.worldbank.org/doi/pdf/10.1596/978-0-8213-6450-5

Anderson, J., & Li, A. (2018). Refugees or victims of human trafficking? The case of migrant domestic workers in Hong Kong. *Anti-Trafficking Review*. https://doi.org/10.14197/atr.201218114

Bai, X., Nath, I., Capon, A., Hasan, N., & Jaron, D. (2012). Health and wellbeing in the changing urban environment: Complex challenges, scientific responses, and the way forward. (Elsevier, Ed.). *Current Opinion in Environmental Sustainability, 4*(4), 465–472. https://doi.org/10.1016/j.cosust.2012.09.009

Branches of Hope. (2024). *5 things you need to know about refugees*. Retrieved from About refugees.

Cap. 115 Immigration Ordinance. (2021). *Hong Kong e-Legislation*. Retrieved from Cap. 115 Immigration Ordinance: https://www.elegislation.gov.hk/hk/cap115

Census and Statistics Department. (2021). *Hong Kong poverty situation report 2020:*. Hong Kong: HKSARG. Retrieved from https://www.censtatd.gov.hk/en/data/stat_report/product/B9XX0005/att/B9XX0005E2020AN20E0100.pdf

Choy, W. C., & Shi, X. (2021, March). A comprehensive review of refugee and asylum studies in contemporary Hong Kong: Law, policies, and lived experiences. *Asian Journal of Social Science, 49*(1), 38–48. https://doi.org/10.1016/j.ajss.2020.11.002

Dahl, R. A. (1961). Who participates in local politics and why. *American Association for the Advancement of Science, 134*(3487), 1340–1348. Retrieved from https://www.jstor.org/stable/pdf/1707610.pdf?refreqid=excelsior%3Ae7d10055908e34226e8875f0a48719be

Equal Opportunities Commission. (2016, July). *Subcommittee on children's rights rights of refugee and asylum seeking children*. Retrieved from LC Paper No. CB(4)1439/16–17(01): https://www.legco.gov.hk/yr16-17/english/hc/sub_com/hs101/papers/hs10120170724cb4-1439-1-e.pdf

Esser, H. (2004). Does the "new" immigration require a "new" theory of intergenerational integration? *International Migration Review, 38*(3), 1126–1159. Retrieved from https://www.jstor.org/stable/pdf/27645428.pdf?refreqid=excelsior%3A9a1fea9cf31bf5811e8f1197fd336eb9

Giddings, B., Hopwood, B., & O'Brien, G. (2002). Environment, economy and society: Fitting them together into sustainable development. *Sustainable Development, 10*, 187–196. https://doi.org/10.1002/sd.199

Grzymala-Kazlowska, A. (2005). From ethnic cooperation to in-group competition: Undocumented polish Workers in Brussels. *Journal of Ethnic and Migration Studies, 31*(4), 675–697. Retrieved from https://web.a.ebscohost.com/ehost/pdfviewer/pdfviewer?vid=1&sid=3c2a7db0-2469-4df7-a2c5-6d6a0e915758%40sdc-v-sessmgr02

Hobfoll, S. E. (2001). The influence of culture, community, and theNested-self in the stress process: AdvancingConservation of resources theory. *Applied Psychology: An International Review, 50*(3), 337–369.

HK Immigration Department. (2021). *Facts and statistics*. Retrieved from Enforcement: https://www.immd.gov.hk/eng/facts/enforcement.html

HK Immigration Department. (2024, September). *Torture/Non-refoulement claim cases*. Retrieved January 27, 2025, from Enforcement: https://www.immd.gov.hk/eng/facts/enforcement.html

HKSARG. (2020, May 20). *LCQ9: Non-refoulement claims*. Retrieved from Press Releases: https://www.info.gov.hk/gia/general/202005/20/P2020052000659.htm

IOM. (2019). *International migration Law: Glossary on migration*. International Organization for Migration.

Justice Centre Hong Kong. (2020, May 5). *Submission to the United Nations human rights committee list of issues on the fourth report of the Hong Kong Special Administrative Region under the international covenant on civil and political rights*. Retrieved July 26, 2021, from Research: https://www.justicecentre.org.hk/framework/uploads/2020/05/Justice-Centre-Hong-Kong-List-of-Issues-Submission-to-Human-Rights-Committee-Final.pdf

Jutvik, K., & Robinson, D. (2020). Permanent or temporary settlement? A study on the short-term effects of residence status on refugees' labour market participation. *Comparative Migration Studies, 8*(44). https://doi.org/10.1186/s40878-020-00203-3

Kellert, S. R. (2012). *Birthright: People and nature in the modern world*. Yale University Press. Retrieved from https://www.jstor.org/stable/j.ctt32bq08

Kennedy, K. J., Wong, K. L., & Ng, H.-Y. (2019). Being asylum seekers in Hong Kong: Institutions, action and power. *Asian Journal of Social Science, 47*, 534–565. https://doi.org/10.1163/15685314-04704006

Kennedy, K., Wong, K. L., & Ng, H. Y. (2016). *Towards a refugee policy for an inclusive Hong Kong: Enhancing the status of China's International City*. The Education University of Hong

Kong, Centre for Governance and Citizenship. Central Policy Unit of the HKSAR Government. Retrieved from https://www.pico.gov.hk/doc/en/research_report(PDF)/2014_A5_010_15B_Final_Report_Prof_Kennedy.pdf

Kuan, T. S., Jiuan, T. S., & Keng, K. A. (2010). *The wellbeing of Singaporeans: Values, lifestyles, satisfaction and quality of life*. World Scientific Publishing Co. Pte. Ltd.

Larsen, B., Kilström, M., & Olme, E. (2018). *Should I stay or must I go? Temporary protection and refugee outcomes*. Department of Economics. Copenhagen Business School. Retrieved from https://research-api.cbs.dk/ws/portalfiles/portal/58519341/WP_5_CBS_2018.pdf

Lau, P. Y., & Gheorghiu, I. (2018, June 7). Vanishing selves under Hong Kong's unified screening mechanism. *Cultural Diversity in China, 3*(1) Retrieved from. https://doi.org/10.1515/cdc-2018-0003/html

Lau, S. (2012, March 29). *Court overturns maid abode ruling*. Retrieved July 6, 2021, from South China morning post: https://www.scmp.com/article/996923/court-overturns-maid-abode-ruling

Legislative Council. (2023). *Panel on security meeting on 5 December 2023 background brief on the handling of non-refoulement claims in Hong Kong*. Legislative Council Secretariat. Retrieved from https://www.legco.gov.hk/yr2023/english/panels/se/papers/se20231205cb2-1078-4-e.pdf

Legislative Council. (2000, February 22). *File ref: L/M 1/99 to SRD 401/1/C*. Retrieved July 6, 2021, from legislative council brief–Vietnamese refugees and migrants: The way forward: https://www.legco.gov.hk/yr99-00/english/panels/se/papers/se-l_m1_99_to_srd_401_1_c-e.pdf

Leung, C., & Lok-kei, S. (2019, January 4). *Tougher measures proposed for overstayers with jobs and employers of illegal workers as Hong Kong government seeks to deter economic migrants*. South China Morning Post. Retrieved from https://www.scmp.com/news/hong-kong/law-and-crime/article/2180625/tougher-measures-proposed-overstayers-jobs-and

Ma, J., & Hamer, L. (2024, March 14). *As more illegal immigrants enter Hong Kong, spotlight shines again on asylum seekers: Do they have valid refugee claims?* Retrieved from South China Morning Post–Law and Crime: https://www.scmp.com/news/hong-kong/law-and-crime/article/3255275/more-illegal-immigrants-enter-hong-kong-spotlight-shines-again-asylum-seekers-do-they-have-valid

Mathews, G. (2018). Asylum seekers as symbols of Hong Kong's non-Chineseness. *China Perspectives, 18*(3), 51–58. https://doi.org/10.4000/chinaperspectives.8132

Matthews, G., Lin, D., & Yang, Y. (2014). How to evade states and slip past Borders: Lessons from traders, Overstayers, and asylum seekers inHong Kong and China. *City & Society, 26*(2), 217–238. https://doi.org/10.1111/ciso.12041

McGranahan, G., Marcotullio, P., Bai, X., Balk, D., Braga, T., Douglas, I., et al. (2005). Urban systems. In *Ecosystems and human Well-being, current state and trends* (Vol. 1, pp. 795–825). IIED. Retrieved from https://www.iied.org/g00505

McGrath, S., & Powell, L. (2016). Skills for sustainable development: Transforming vocational education and training beyond 2015. *International Journal of Educational Development, 50*, 12–19. https://doi.org/10.1016/j.ijedudev.2016.05.006

Ng, S. M., Law, T., & Yeung, T. (2019). *Study on the discrimination experience based on residency status of asylum seeker/refugee in Hong Kong–research report*. The University of Hong Kong & Equal Opportunities Commission. Retrieved from https://651001ff-6ce1-4a42-bc36-956cf637f9bd.filesusr.com/ugd/05b9cb_ea9909c0476f4309bdee4313fe2c6bb0.pdf

Norman, K. P. (2017). *Reluctant reception: Understanding migration and refugee policy in Egypt, Morocco and Turkey (DISSERTATION)*. Department of Political Science. University of California. Retrieved from https://escholarship.org/uc/item/4t07g3rh#main

Norman, K. P. (2018). Migration and refugee policy-making in modern Egypt, Morocco and Turkey. *Mashriq & Mahjar: Journal of Middle East and North African Migration Studies, 5*(2) Retrieved from https://muse.jhu.edu/article/778335/pdf

Norman, K. P. (2019). Inclusion, exclusion or indifference? Redefining migrant and refugee host state engagement options in Mediterranean 'transit' countries. *Journal of Ethnic and Migration Studies, 45*(1), 42–60.

Nussbaum, M. C. (2000). Introduction: Feminism and international development. In M. C. Nussbaum (Ed.), *Women and human development: The capabilities approach* (pp. 1–31). Cambridge University Press.

Ramsden, M., & Marsh, L. (2013). The 'right to work' of refugees in Hong Kong: MA v director of immigration. *Internatinoal Journal of Refugee Law, 25*(3), 574–596.

Refugee Concern Network. (2021, February). *Submissions to the bills committee on immigration (Amendment) Bill* 2020. Retrieved July 26, 2021, from Research: https://www.justicecentre.org.hk/framework/uploads/2021/02/Refugee-Concern-Network-Submissions-on-the-Immigration-Amendment-Bill-2020.pdf

Riedel, B. (2017, December 18). *Who are the Houthis, and why are we at war with them?* Retrieved from Brookings: https://www.brookings.edu/blog/markaz/2017/12/18/who-are-the-houthis-and-why-are-we-at-war-with-them/

Ryan, D., Dooley, B., & Benson, C. (2008). Theoretical perspectives on post-migration adaptation and psychological well-being among refugees: Towards a resource-based model. *Journal of Refugee Studies, 21*(1), 1–18.

Ryan, R. M., & Deci, E. L. (2001). On happiness and human potentials: A review of research on hedonic and Eudaimonic Well-being. *Annual Review of Psychology, 52*, 141–166.

Sadr-Mohammadi, R., Kalantari, M., & Molavi, H. (2014). Efficacy of life skills training on subjective well-being of students: A report from Rafsanjan, Iran. *Iranian Journal of Psychiatry and Behavioral Sciences, 8*(2), 63–67. Retrieved from https://www.ncbi.nlm.nih.gov/pmc/articles/PMC4105605/

Save the Chilren Hong Kong et al. (2017, July 24). Joint submission to subcommittee on children's rights meeting on 24 July 2017 re: Rights of refugee children. Retrieved from Legislative Council: https://www.legco.gov.hk/yr16-17/english/hc/sub_com/hs101/papers/hs10120170724cb4-1458-2-e.pdf

SCMP. (2012). *Four refugees, torture claimant seek ruling on right to work in Hong Kong*. Retrieved from https://www.scmp.com/news/hong-kong/article/1029583/four-refugees-torture-claimant-seek-ruling-right-work-hong-kong

Secretary for Security. (2021). *Replies to initial written questions raised by finance committee members in examining the estimates of expenditure 2021–22*. Security Bureau Finance Committee. Retrieved from https://www.legco.gov.hk/yr20-21/english/fc/fc/w_q/sb-e.pdf

Security Bureau. (2018, April 24). *Subcommittee to follow up issues relating to the unified screening mechanism for non-refoulement claims pre-arrival control*. Retrieved from Legislative Council: https://www.legco.gov.hk/yr16-17/english/hc/sub_com/hs54/papers/hs5420180424cb2-1250-1-e.pdf

Sen, A. (1999). *Development as freedom*. Oxford University Press.

Stephenson, J. (1998). The concept of capability and its importance in higher education. In J. Stephenson & M. Yorke (Eds.), *Capability and Quality in Higher Education* (1st ed., pp. 1–13). Routledge. https://doi.org/10.4324/9781315042046

Shum, T. C. (2014). Refugees' transnational mobility: A study of asylum seeking in Hong Kong and urban Thailand. *Refugee Survey Quarterly, 33*(4), 50–80. Retrieved from https://www.jstor.org/stable/pdf/45054723.pdf?refreqid=excelsior%3Afba8f52f0ac0c370c02275ea938f3c86

Szekely, E., & Mason, M. (2019). Complexity theory, the capability approach, and the sustainability of development initiatives in education. *Journal of Education Policy, 34*(5), 669–685.

Thomas, J. (2000). *Ethnocide: A cultural narrative of refugee Detentin in Hong Kong*. Ashgate Publishing Ltd.

Ungera, J. M., Rauchb, A., Fresec, M., & Rosenbusch, N. (2009). Human capital and entrepreneurial success: A meta-analytical review. *Journal of Business Venturing, (26)*, 341–358. https://doi.org/10.1016/j.jbusvent.2009.09.004

van Agteren, J., Iasiello, M., & Lo, L. (2018). Improving the wellbeing and resilience of health services staff via psychological skills training. *BMC Research Notes, 11*, 924. https://doi.org/10.1186/s13104-018-4034-x

Vecchio, F. (2016). *Asylum seeking and the Global City*. Routledge.

Veenhoven, R. (2008). Sociological theories of subjective Well-being. In M. Eid & R. Larsen (Eds.), *The science of subjective Well-being: A tribute to Ed Diener* (pp. 44–61). Guilford Publications.

Wells, T. (2012). *Sen's capability approach*. Retrieved from Internet Encyclopedia of Philosophy: https://www.iep.utm.edu/sen-cap/

Western, M., & Tomaszewski, W. (2016). Subjective wellbeing, objective wellbeing and inequality in Australia. *PLoS One, 11*(10).

Wong, W. C., Cheung, S., Miu, H., Chen, J., Loper, K. A., & Holroyd, E. (2017, February 2). Mental health of African asylum-seekers and refugees in Hong Kong: Using the social determinants of health framework. *BMC Public Health, 17*(153). https://doi.org/10.1186/s12889-016-3953-5

Chapter 3
Empowerment of Refugees for Greening Societies and Economies: Approaches for Capacity Building Through Green Skills Development

Abstract Migration can increase the burden on the host cities regarding the need for additional development, as outlined in the Global Compact on Refugees (United Nations, *The future we want*. Retrieved from UN Sustainable Development, 2012). With an increase in the number of people in a particular region/area, economic activities, the use of natural resources, (urban) infrastructure operations, equal access and distribution of food and water, as well as essential services such as accommodation, health, and quality of education are affected. Each of these elements contributes to achieving sustainable development of a city. As Brown (*Migration and climate change*. International Organization for Migration, 2008) argues, in the short term, forced migration due to climate change can influence the achievement of SDGs. In the long term, climate change-forced migration (such as migration due to sea-level rises) can roll back the progress that has been achieved so far as it interrupts the provision of education and health-related services, both of which relate to several SDGs such as universal primary education (SDG 2) and reducing child and maternal mortality, as well as combating HIV/AIDS, malaria and other diseases (SDG 4 and 5) (Brown, *Migration and climate change*. International Organization for Migration, 2008). Therefore, regardless of whether migration is caused by climate change or other issues, the hosting cities need to ensure that migrants can contribute to their sustainability efforts. Therefore, a focus should be on capacity development, as many migrants have limited access to resources and might feel powerless.

Therefore, this chapter links concepts related to sustainable urban development and refugees, such as empowerment, challenges before achieving sustainability, the need for skills development, especially for refugees and asylum seekers, and social entrepreneurship as an avenue for empowerment.

Keywords Empowerment for refugees · Generic green skills · Social entrepreneurship for refugees · ESD pedagogy

A. T. Saral, M. Pavlova, *Education for Refugees and Sustainable Green Urban Development*, Education in the Asia-Pacific Region: Issues, Concerns and Prospects 78, https://doi.org/10.1007/978-981-95-1680-3_3

3.1 Empowerment

Alsop et al. (2006) define empowerment as a group's or an individual's capacity to make effective choices and then transform those choices into desired actions and outcomes. Therefore, it can be conceived of as both a process and a product. However, some factors influence this capacity, and these are the opportunity structure and agency (discussed in Chap. 2). The opportunity structure refers to rules and formal and informal institutions that shape and constrain human interaction and behavior. On the other hand, an agency is about individuals or groups making choices as civic, economic, or social agents. The agency is related to an individual's resources, such as psychological, informational, and material resources, as well as organizational, social, financial, and human resources (Alsop et al., 2006). As Sen (1999) argued, resources are essential to understanding a person's capability and assessing well-being. Ryan et al. (2008) classified these resources as personal, social, cultural, and material (discussed in detail in Chap. 2) in relation to refugees' lives. In the study we are presenting in this book, having access to resources by RAS is directly related to empowerment as it indicates an ability to choose from different options.

The literature demonstrates that even if a person has the capacity and resources, they may not be able to transform their agency into action due to available opportunity structures. This situation can influence the integration of stigmatized groups into societies. From this perspective, Esser (2004) demonstrates that the integration or disintegration of groups within a society is a structural outcome. Agents' reactions shape these structures into societal conditions and vice versa. According to Esser (2004), in all immigration processes, there is a structural change in institutional and cultural conditions for productive actions, meaning that specific resources and investments are more efficient than others for attaining the respective (cultural) goals. Some institutional and cultural core elements influence new immigrants' orientation in their new setting, so resources brought from the homeland and capital are valuable but may need more usability and productivity according to the needs of the new settlement. Therefore, more generalized forms of capital, such as technical and administrative knowledge, would free people from such limits and be useful for integration and empowerment. Certain areas need to be considered to understand and track empowerment. One of the areas that needs to be clarified is the degree of empowerment:

1st degree—Existence of choice: whether an opportunity to make a choice exists or not

2nd degree—Use of choice: whether a group or a person uses the opportunity to make a choice

3rd degree—Achievement of choice: whether choice brings about the desired result (Esser, 2004).

Although degrees of empowerment, such as first degree and second degree, seem too mechanical and lead to another, this is not always the case, according to Esser (2004).

For example, Hong Kong asylum seekers may want to participate in a particular training program, such as producing solar panels, which is relevant to the greening agenda. The opportunity to choose exists if training is provided by public or private means. If not, there is a need to provide opportunities for people to be empowered to the first degree. When the training is offered, the person can choose to participate. If the person decides on no participation, then the opportunity structures and agency must be interrogated. Did financial resources hinder participation, or did the choice relate to informal institutions, such as a lack of self-esteem that prevented the person from participating? Therefore, each context should be analyzed in association with agency and opportunity structure. For example, a study in Nepal showed that excluding women is both the cause and the result of unequal development. Although there are progressive steps towards bridging the gender gap at the policy, legal, and programming levels—formal institutions –attitude problems can remain in society due to norms and values—informal institutions. Social exclusion is argued to be both a structural and a social problem. Tackling such structural issues requires an intervention approach, an action for change, and inclusive development, such as a change in policy, the introduction of programs, and the implementation of these initiatives. From this perspective, access to skills training, education for increasing awareness, advocacy work, and diversity incentives are key factors (Asian Development Bank, 2020).

Another area is gender mainstreaming, which is defined as

> ...the process of assessing the implications for women and men of any planned action, including legislation, policies or programmes, in all areas and at all levels. It is a strategy for making women's as well as men's concerns and experiences an integral dimension of the design, implementation, monitoring and evaluation of policies and programmes in all political, economic and societal spheres so that women and men benefit equally and inequality is not perpetuated. The ultimate goal is to achieve gender equality… Gender mainstreaming does not replace the need for targeted, women-specific policies and programmes or positive legislation, nor does it substitute for gender units or gender focal points. (UN Women, 2020, p. 2).

It is argued that gender mainstreaming is the most practical and inclusive means to achieve gender equality and women empowerment (UN Women, 2020). Becker (1995) observes that women refugees and minority groups underinvest in their human capital because they have been raised to believe they cannot do certain things, an example of an informal institution (cited in Alsop et al., 2006). Nussbaum also argues that perpetuating these roles leads to the internalization of disempowerment and choice-making without considering other possibilities (Nussbaum, 2000).

3.2 Sustainable Development and Greening Economy: The Need for Skills

Hong Kong positions itself as the "World City of Asia" thanks to its highly competitive economic status, international connections, and education services focusing on knowledge creation and enhanced quality of life. However, Hong Kong's high economic status does not guarantee social inclusion, influencing sustainable development efforts negatively. HKSAR has implemented some actions to address three sustainability areas: economy, society, and environment.

The economic meaning of sustainable development is about being fair to the future and getting accounting systems to reflect, as far as possible, on the shadow prices of the environment[1] (Pearce, 1989). For sustainable economic growth, besides considering the environmental impact and limited resources, technological and innovative capacity is a key indicator of economic growth and development. Hong Kong SAR developed policies and established institutions to promote innovation as a driver of sustainable economic growth (e.g., Policy address, 2015 cited in Pavlova & Askerud, 2023). Innovation is viewed as improving livelihood opportunities and quality of life—notably for people experiencing poverty—through reduced environmental impacts, improved use of natural resources, and increased and more equally distributed GDP (UNEP, 2024). However, implementing these policies depends on the availability of talents with the required skills.

A report by LinkedIn Economic Growth shows that between 2018 and 2023, the share of green talent grew by 5.4% per year, while the share of jobs requiring at least one green skill grew by 9.2%. However, seven in eight workers worldwide lack even a single green skill (LinkedIn Economic Graph, 2023). The problem behind this mismatch was found to stem from:

(i) an inability to meet new requirements of emerging (green) jobs,
(ii) a shortage of skills in high-demand green jobs such as renewable energy.
(iii) the poor reputation or unattractiveness of specific sectors, such as waste management (International Labour Organization, 2013).

This situation hampers countries' green investments and growth and emphasizes the necessity of retraining in green skills (MacLean, 2018a, 2018b).

Though the HK government has introduced multiple initiatives that serve to green the economy and achieve SUD, there are still environmental, institutional, and social constraints to overcome.

According to statistics, the gross domestic expenditure on research and development (GERD) in Hong Kong in 2022 increased by 8% compared to the previous year and totaled $30,138 million. In terms of spending on innovation activities by the business sectors in 2022, the information and communications sector accounted for the largest share (46%), followed by the financing and insurance, real estate, professional, and business services sectors (23%); and import/export, wholesale and retail trades, and accommodation and food services sector (22%). Analyzed by type of innovation activity, the total expenditure of innovation activities in the business sector in 2022 was mainly engaged in in-house R&D activities (such as natural sciences, engineering and technology, medical, health, social sciences, and humanities

[1] The shadow prices of the environment relate to the lack of reflection of the environmental services and the damage in the pricing in the market economy (for example, if timber is removed unselectively from a rain forest, there is damage to the watershed through river pollution and soil erosion. However, these environmental costs are not considered in the price of timber (Pearce, 1989, p. 5).

and arts) conducted for own establishment (31%), followed by R&D activities contracted out to other parties (23%); and engineering, design, and other creative work activities (17%) (HK Census and Statistics Department, 2023). The way the government collected these statistics is unclear as to what percentage of 'green' or 'sustainable' innovation is in place. Thus, the background of personnel employed in such activities and requirements for green skills is not transparent. However, this can indicate that there is no particular focus on green innovation as requested by the government.

The unstable domestic market demand, the insufficient build-up for small and medium-sized enterprises (SMEs) and start-ups (where innovative and technological activities are undertaken), as well as the human resources capacity/skills gap to meet the demand for innovation and environmental solutions in the industry (Choi, 2024) are among the identified challenges, indicating the need for and importance of appropriate skills development. Although policies for greening are in place in Hong Kong, there is a lack of systematic definition and identification of those skills needed for greening the economy (Pavlova, 2019). She argues that the disconnect between policy and action may reflect a lack of understanding and concern for transitioning into greening. Pavlova (2019) suggests

(i) the implementation of efficient and inclusive models of new skills supported by technologies, policies, and training inclusively.
(ii) further integration of green skills in the mainstream industry (Pavlova, 2019).

In terms of environmental sustainability, legal frameworks, laws, and regulations at the state level are highly influential in ensuring the sustainability of the environment in urban areas, steering people to act in specific ways, and leading the way for technological advancement (Vona, Marin, Consoli, & Popp, 2015). In this regard, rapid urbanization in Hong Kong has brought about multiple environmental issues related to a need for energy efficiency, waste management, and air and water quality (they were discussed in Chap. 1). Although these initiatives are very important, some of them were not successful. For example, in 2020, plastics accounted for 21% of the city's total municipal solid waste (MSW), which is the third-largest share of MSW after food waste and paper. The city's beaches and waterways are drowning in plastic, and microplastic levels in the sea are 40% higher than the global average. The government put forth the MSW charging scheme in 2024; however, upon consideration and public views, the government decided to suspend it. Instead, to make citizens accept and adapt to it, the food waste recycling network has been expanded, the operating hours and the number of collection points of GREEN@COMMUNITY recycling facilities have increased, and additional recycling facilities have been installed for people and companies (Environmental Protection Department, 2024). Though at the enterprise level, it is observed that some food and beverage companies have tried to adopt biodegradable and even compostable takeaway packaging and utensils, most local restaurants and small businesses still opt for low-cost materials like Styrofoam. In addition, a survey by the World Green Organisation found that Hong Kong residents lacked awareness of sustainability (Leung, 2019). These results show that an efficient recycling infrastructure is not enough to overcome the plastic crisis in Hong Kong (Lai, 2024); the level of people's awareness concerning sustainability also needs to be improved. For this reason, reflection on how to ensure economic growth without compromising environmental resources, developing the capacity for SD innovation and ensuring inclusion should be a forward-looking task in Hong Kong and beyond.

Regarding social sustainability, socially excluded groups are often at the bottom of the city's social hierarchies. Urban renewal policies, such as redevelopment and rehabilitation, allow disadvantaged residents to improve their housing conditions and lower their transportation costs, thus contributing to social integration and interaction and promoting social cohesion and sustainability (Ho et al., 2012). To address social sustainability, HK adopted a "People First, District-based, Public Participatory" approach that emphasizes public engagement in its policy development for sustainability (Chan & Siu, 2015). However, studies concerning disadvantaged groups living in HK repeat the feeling of exclusion and unsatisfactory attainment due to lack of access to education, health, and employment services as compared to locals, indicating the difference between intentions and outcomes (Kennedy et al., 2017; Bhowmik, 2014; Lee & Law, 2016). Another social constraint for achieving sustainable development in HK is the high aspiration of the working-class population in the city for economic survival, which weakens their desire for greening the city and sustainability (Tian, Jim, and Tao, 2012). Although they might care about the environment, people are too busy to consider the impact of their actions on the environment and others, indicating a discrepancy between values and actions.

These constraints prove Pavlova's (2022) argument that environmentally conscious values and actions could not be developed separately because a weak relationship between them means that when the situations are ameliorated, or there is discomfort in conducting such behavior, the person does not continue it (Arbuthnott, 2009). Thus, she suggests that developing generic green skills is an enabling factor, creating positive values that can be transformed into the required behavior. In addition, Vona et al. (2015) also argue that green skills are a set of competencies related to the design, production, management, and monitoring of technology in the context of environmentally sustainable development (Vona, Marin, Consoli, & Popp, 2015)—not only understanding the meaning of the word sustainable but also pragmatic learning. For that, an approach to SUD requires organizational regulations at the state level and a human force that can understand sustainability issues, implement them in their lives, and work with innovative technologies and upcoming trends on the individual level. There is a need for coordination and partnership at all levels (Pavlova, 2019).

The European Centre for the Development of Vocational Training (CEDEFOP) suggests that developing a low-carbon economy depends on improving existing skills rather than developing specialized green skills (CEDEFOP, 2010). We also believe that generic skills are fundamental, as they are the core competencies that can be transferred and built upon. "*…By breaking down roles into the specific capabilities required to do them, we can develop talent strategies that recognize individuals for the capabilities they possess. By thinking of climate-related jobs as* *collections of skills, specifically "green skills," we can expand the talent pool available to solve the climate crisis"*. (LinkedIn Economic Graph, 2023). Moreover, include refugees and asylum seekers in the green agenda through training.

3.3 Generic Green Skills

It is argued that the core green skills to be delivered via education are the willingness to learn about sustainable development, increased awareness of environmental issues, and general learning and decision-making ability related to consumption, production, waste, and energy. Pavlova calls these core abilities *generic green skills* (Pavlova, 2018) and argues that the development of generic skills can be expected to lead to 'economically viable, environmentally sound and sustainable communities' (Paryono, 2017). A learning culture is required to position the learner in the center and connect with sustainability issues. As such, generic green skills foster learning about sustainability by focusing on four aspects, as indicated in Diagram 3.1:

Generic green skills (GGS) are a slightly newer term in the literature compared to sustainability education, education for sustainable development (ESD), and environmental education (EE), which date back to the 1987 Brundtland Report and 1992 Earth Summit (Brundtland Commission, 1987). Despite being defined in various ways and incorporated into different subjects in the curricula in formal learning settings, the terms derive from a similar objective: ensuring sustainable development. For example, in Australia, they are called *skills for sustainability*. They refer to technical skills, knowledge, values, and attitudes needed in the workforce to develop and support sustainable social, economic, and environmental outcomes in business,

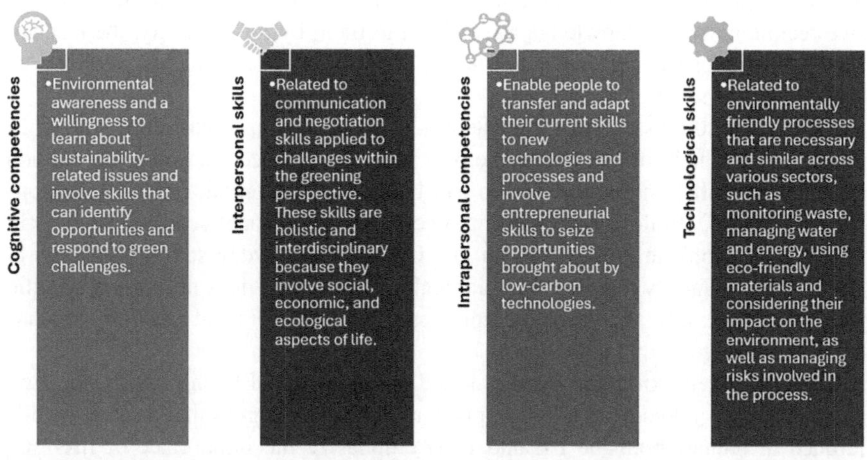

Diagram 3.1 Generic green skills. Source: Pavlova, 2019

industry, and the community (Maclean, 2018a, 2018b). The Organization for Economic Co-operation and Development (OECD) classifies green skills together with entrepreneurial skills under converging skills, meaning that these are the specific skills required to adapt products, services, or operations to meet adjustments related to the impacts of climate change (Maclean, 2018a, 2018b). In Indonesia, however, some (TVET) providers have considered green skills related to hygiene and safety, while others regarded it as a new curriculum related to environmental standards (Maclean, 2018a, 2018b).

In Hong Kong, these skills are under environmental education and are coordinated through cooperation between departments, such as the Education Bureau, Environment Bureau, and the Council for Sustainable Development Education (Education Bureau, 2010), and are supported by (i) school inspections and visits, (ii) teacher education programs, (iii) provision of learning resources and materials, and (iv) organization and funding of territory-wide activities (Education Bureau, 2016). The Education Bureau has identified crucial areas of sustainable development and has included the themes identified below via a cross-curricular approach at different levels. At the primary level, the themes are incorporated into the General Studies subject; at the secondary level, they are integrated into Geography, Science, and Technology Education; at the senior secondary level, under Geography, Biology, and Liberal Studies (Education Bureau, 2010; Ma, 2019). Some of the themes covered in these subjects are Climate change and global warming, green school campuses, urbanization and housing problems, waste reduction, and recycling (see pp. 8–10, Education Bureau, 2010 for more detailed information about the subjects and themes at each level).

Pavlova and Chen (2019) observed TVET classrooms in Hong Kong. They concluded that the lectures were content- and teacher-centered and did not consider the complexity and connectedness of sustainability issues or lead students to explore real-world problems. This is a traditional education setting, where learners are passive recipients of the knowledge transmitted to them by the authority, the teacher. Therefore, the process creates a disconnect between knowledge and its application, which requires skills.

Chisholm (2005) states that knowledge should be combined with skills and competencies to be helpful and with principles and values to become worthwhile. Such knowledge, skills, and competencies can be learned in formal, informal, or non-formal settings. While formal learning occurs in institutional settings (such as in school education), informal learning can happen in everyday settings, such as at work or in the family (such as in incidental learning), and does not have a specific purpose. Conversely, non-formal learning can be intermittent and transitory and has a particular target group and purpose.

Generic green skills can be developed through both EE and ESD education within general and vocational programs. Pavlova (2012) argues that despite the difference in names, both the EE and ESD emphasize the importance of life-long learning, the integration of social, environmental, and economic realms (sustainability), the use of a variety of techniques that promote participatory learning, first-hand learning, and the development of higher-order thinking skills such as problem-solving and critical thinking, therefore enabling the application of knowledge and skills in real life. Whether they are EE or ESD, in formal or non-formal settings, the focus should be on understanding local and global links, relating learning to the community, and addressing SD concepts that deal with socio-ecological structures and capacity building for communities.

The above indicates that implementing environmental or sustainability education to develop generic green skills is complex and varies across contexts.

3.3.1 Settings for Learning Green Skills

The lack of a clear and standard definition of green skills may limit the effectiveness of the education and training provided. The lack of definition and variations stem from the differences in social, economic, and research needed and lead to problems in measuring good practices and identification of demand and supply (Rikala, Braun, Järvinen, Stahre, & Hämäläinen, 2024). However, the common ground is that the desired achievement/attainment of a learning activity is the application of one's knowledge to a problem or transferring this ability between different situations/tasks at the end of the learning process (Chisholm, 2005). If the skills or knowledge are not transformed into desired actions (in the context of this book, it is RAS's care for the environment and their conscious environmental action), then is that knowledge worth knowing or valuable? Stevenson (2006) argues that one cannot address a vision of a sustainable society without simultaneously considering the present environmental issues that need to be tackled. So, he suggests that to transform society and the environment, creating a space where educators are involved in the discourse is necessary. It should be informed historically, pedagogically, and politically by practice in diverse cultural contexts. Regarding *formal learning*, Ma (2019) criticizes the non-compulsory status of environmental education in Hong Kong and that implementation is left to schools' discretion. She suggests adopting a threefold approach to EE: "Education in/for/about Environment." She states that Hong Kong's EE lacks "in" and "for" aspects of environmental education. Education "in" the environment can be considered an outdoor learning activity, where the learner places themselves in nature and develops environmental sensitivity—which responds to features of non-formal learning, for it can take place outside school and be a result of informal learning because the motivation belongs to the learner -. Education "for" the environment stresses developing responsibility for acting and solving environmental and development problems—which is an expected result for any learning as it transforms into behavior. However, the government does not indicate strategies and policies for involving all community members and increasing public awareness. Thus, it would be challenging to foster the idea and active participation (Ma, 2019). Another study conducted in Hong Kong concurs with this notion. It reveals that although there was a perceived increase in knowledge of selected environmental issues in schools, approximately 30–50% of students reflected no behavioral change (Maclean & Pavlova, 2017).

As argued by Mok (2011) education cannot be confined to time-bound and space-bound school settings or quantified by years of exposure and concluded that education could be equated to learning (Mok, 2011). Thus, today, people learn in different settings such as bookstores, video stores, museums, cyberspace, summer camps, retreats, and theme parks. The term "edutainment" has been coined to refer to entertainment that educates. Many of these new modes of education have been collectively denoted "informal education." (Chazan, 2003).

It is not easy to cover the full extent of informal learning opportunities in which people engage throughout their lives because, as Yang (2015) argues, informal learning, by its nature, can be intentional and often fostered in public areas, including TVs in metros, signboards, or cultural/religious centers (UIL, 2022).

Sustainable urban development requires structuring public space to ensure informal learning. An example of this is the Kai Tak Sports Park in Hong Kong. The government has incorporated green buildings and ecological designs, renewable energy systems, and innovative city concepts into the park (Kai Tak Sports Park Ltd., 2024), exposing residents and visitors to green solutions and developments.

Non-formal learning is purposive, but voluntary learning takes place in a diverse range of environments and situations. For this reason, teaching/training and learning is not necessarily the sole or primary activity. The activities are planned but are seldom structured by conventional curriculum subjects. They usually address specific target groups but rarely assess the learning outcomes in a traditionally visible way (Chisholm, 2005). Hence, it supports informal and formal learning.

The characteristics of each type of learning are included in Table 3.1. Learning relevant to non-formal learning is bold.

Each type of learning has its advantages and disadvantages. For example, in comparison with other learning types, formal learning has well-established methodologies for assessment and evaluation. However, there are some significant dropout rates in formal teaching programs (Powdyel, 2016) or no courses offered, as in the case of refugees and asylum seekers in HK. Therefore, non-formal platforms are helpful to compensate for the limitations of formal learning and encourage social inclusion by targeting specific, marginalized groups (Johnson & Majewska, 2022). Sometimes, a lack of formal assessment is considered a disadvantage of non-formal learning that can bring it a lower status (Johnson & Majewska, 2022). However, non-formal education can complement formal education by fostering the development of skills and competencies, for example, by collaborating with different institutes and promoting additional sets of skills.

In terms of teaching, educational institutions have traditionally been far more focused on effective individual development and a particular way of teaching (that stresses the transmission of knowledge and teacher-centered side) than fostering collaborative skills and the collective agency needed by communities and social movements. So, if we ask teachers to change how and what they teach, it should change the organizational and political context (Stevenson, 2007). Moreover, even if it is politically achieved, negotiating and enacting changes in practice is much more complicated than negotiating a shift in policy and rhetoric because the former challenges vested interests in the status quo. In addition, the amount of space

Table 3.1 Types of learning

Formal Learning	Non-formal Learning	Informal Learning
Learning is structured (e.g., linear objectives)	**Learning may be structured**	Learning is not structured
Learning is promoted through direct teaching behaviors	**Learning is promoted through indirect teaching behaviors**	
Learning is intended (by educator and learner)	**Learning is intended by the learner**	Learning may not be intended by the learner
Learning is recognized by the learner and educator	**Learning is recognized by the learner**	Learning may not be recognized by the learner
Motivation for learning may be extrinsic to the learner		Motivation for learning is intrinsic to the learner
Learning takes place in educational institutions	**Learning can take place in educational institutions**	Learning can take place anywhere
Learning has a mandated dimension	**Learning has a voluntary dimension**	
Learning may be recognized or measured through qualifications		Learning is not recognized or measured through qualifications
Learning may primarily focus on propositional knowledge	**Learning may focus on both propositional and procedural knowledge***	
Learning tends to have a cognitive emphasis	**Learning involves cognitive, emotional, social, and behavioral elements**	

**Propositional knowledge includes conceptual knowledge (e.g., 'knowledge-that') and contrasts with procedural knowledge which includes competencies, techniques, and skills (e.g., 'knowledge-how') (Rata, 2019).*

Source: Johnson & Majewska, 2022

devoted to participation, self-determination, and independent thinking has implications for how education for sustainability (ESD) is perceived and realized. It is reported that when this space is narrow, teaching and learning practices tend to be more transmissive and teacher centered. Although this approach may be advantageous in reaching more people and achieving a desirable result, such as behavioral change about sustainability, it does not promote a meaningful engagement/internalization with the subject. However, ESD tends to prevail when the space for participation is broad, more interactive, student-centered, and transformative. It has been argued that this mode of teaching and learning leads to capacity building, empowerment, and behavior change (Shaping the Education of Tomorrow, 2012).

In the case of the refugee/asylum-seeking community and their children in HK, the latter can attend schools and learn about environmental issues there. However, training and education are often limited regarding what NGOs can offer adults. No systematic environmental education is available to parents with a claimant background in Hong Kong. Therefore, children's practices, attitudes, and learning are more challenging to support. Hence, it is necessary to support education and offer a chance to learn about sustainability in informal or non-formal settings.

Formal or non-formal educational settings are composed of learning and teaching experiences, which define a pedagogy and one's teaching stance and require an understanding of how learning occurs. Regarding educational objectives, Bloom's taxonomy often sets the blueprint and has been highly influential in developing pedagogy (Blooms, Engelheart, Edward, Walker & Krathwol, 1956). *Cognitive areas* include remembering, recalling knowledge, thinking, problem-solving, and creating. The *affective domain* relates to emotions and describes a change in interests, attitudes, values, and the development of appreciation. The *psycho-motor domain* is concerned with motor skills or skilled behavior. The relationship between these domains is intricate because cognition can occur with a minimum of motor activity, and the affective domain includes a considerable amount of cognition and emotion/feeling. As the name suggests, psycho-motor involves cognition and affection in the willingness to act (Simpson, 1966). From this perspective, Sern et al. (2018) observe (generic) that green skills cover all three domains of learning, all of which are necessary for learners (in their article, they are 'workers') to promote sustainable development in the social, economic, and environmental domains (Sern, Zaime and Lee, 2018). Therefore, if we apply Bloom's taxonomy in the classroom where generic green skills are taught, each domain should be covered to ensure an understanding of SD-related issues. For example, when learning about waste management, issues about how it applies to daily life and industry, the challenges faced, and responses to these problems are at the core of developing cognitive competencies of generic green skills. In terms of the affective domain, realizing one's disposition towards the environment and waste production (such as eco-centric or anthropo-centric orientation) and developing a sense of belonging to nature as well as communicating effectively in conflicting issues that arise from this issue fall under this domain, and are covered by interpersonal competencies in generic green skills. In terms of the psycho-motor domain, the development of technical skills is necessary for the production of renewable energy sources, transferring and topping up current skills in a new job, as well as turning waste into a green-friendly opportunity/action such as in social entrepreneurship (covered under intrapersonal competencies). Sen (1999) argues that the sustainability of development depends on people's role as agents of change. Furthermore, the role of education systems is to develop the capabilities of current and future generations to achieve sustainable livelihoods and well-being within peaceful and democratic societies. This requires considering disadvantaged groups' institutional barriers to realizing valued capabilities and functioning (Tikly, 2019).

In this context, the development of generic green skills through non-formal learning aims for social transformation, which values human beings at its core and supports the greening of the economy; it values the environment but also acknowledges the importance of economic activity for sustaining our lives because non-formal learning allows interaction between education, public awareness, and training (Shaping the Education of Tomorrow, 2012). Hence, education for sustainable development through non-formal settings is essential to bridge the gap between knowledge and behavior. To achieve SUD, it is imperative that all people, including refugees and asylum seekers, be valued and included in environmental education.

3.3.2 *Skills Development for Refugees*

Studies about skills development (for disadvantaged communities), in general, highlight their importance for integrating these groups into society. International organizations' reports (Strietska-Ilina et al., 2011; CEDEFOP, 2010) conclude that short, intensive courses for skills development are worth delivering and instrumental in improving the quality of disadvantaged people's lives (SDG 3) and for providing inclusive, sustainable growth (SDG 8).

NGOs usually play a crucial role in that. In Ethiopia's Somali and Tigray regions, refugees are provided short-term training in an ad-hoc manner with NGO sponsorship for about 2 weeks, and women's participation rate is poor (Tekleselassie, 2021). In these regions, local NGOs are crucial in providing skills training for refugees, and international organizations support them. For instance, a project in Kakuma camps and Turkana West in Kenya, funded by GIZ (the German Development agency), was reported to provide information and communication technology (ICT) courses to refugees to improve their prospects for the future (GIZ, 2019). The UNHCR-funded project in Pakistan provided beautician, tailoring, and dressmaking courses via four National Vocational and Technical Training Commission Centers in different cities. The aim was designed to enable refugees to support themselves in Pakistan. It also helped better prepare refugees to contribute to reconstructing their country if they decided to return to Afghanistan (Karim, 2021). Likewise, Zaatari Camp in Jordan offered tailoring, hairdressing, English/French language, drawing, and craft-making courses. As a result, women refugees reported improved confidence and self-esteem (Jabbar & Zaza, 2016). These types of training, such as carpentry and hairdressing, are often found in skills training for refugees. The training usually aims to find short- or mid-term solutions to refugees' employment problems and enable them to earn money. Therefore, the focus is generally on immediate economic and social survival rather than sustainability (environment, social, or financial). There are examples of green skills development for refugees. In Kenya the development of skills among refugees in the fields of sustainable energy production, construction, and management that were applied to build a new settlement for them in Kenya resulted in reduced hostility between refugees and host groups and the creation of jobs for refugee men and women (SDG 8) (UN Habitat, 2018). Another study evaluating a skills development program for poverty-affected and refugee women in Jordan revealed that skills development significantly contributes to well-being and personal development (SDG 3) (Thorne, 2020). Similarly, the results of a European Union project called URBACT demonstrate the importance of including excluded groups, such as migrants and minorities, for socially inclusive and sustainable urban development efforts (SDG 11) and suggest that building partnerships are essential for these groups to have a better quality of life (URBACT, n.d.). However, opportunities for skills training and access to work remain a challenge for disadvantaged and marginalized groups, like refugees and asylum seekers (SDG 4) (King & Palmer, 2010). One associated problem regarding this matter is the recognition and relevance of refugees' skills in the country of arrival. However, Paryono (2017) argues that recognition and skills transfer are essential for mobility and migration within and across countries. A report on migration and its impact on cities emphasizes that one in seven people on the planet is from a migrant background, including refugees and asylum seekers, and that they have a skills spectrum ranging from unskilled to highly skilled (World Economic Forum, 2017). Considering that more jobs will be available in many areas as countries adopt a cleaner economy, it is essential to integrate the refugee-like demographic into this force to stimulate green development (UNICEF, 2024). For example, over 20 million new clean energy jobs are expected to be created by 2030 (IRENA, 2022), and jobs in construction and operations are expected to grow by 55 million (Huckstep & Kenny, 2022). A Latin America and the Caribbean report also shows that other industries, like farming, forestry, and manufacturing, expect millions of new green jobs (Saget et al., 2020). Moreover, if not green jobs, many hospitality, tourism, healthcare, education, and human services jobs will be required to become more eco-friendly. Therefore, it is essential to integrate this demographic force to stimulate diversity, entrepreneurship, and prosperity in their host cities by removing barriers to employment, education, and urban resources and goods. These are key factors for sustainable urbanization (UN-Habitat, 2017).

Among refugees and asylum seekers, women are particularly vulnerable. They are often discriminated against and are physically susceptible to danger. They also lack equal access to funds than men or participate in traditional "women's work" such as sewing and cooking (Trier, 1994). Refugees have ambitions, desires, and hopes. They are driven first by self-preservation but expect to earn a living and to be able to reconstruct themselves in the long term instead of living on welfare benefits. Several studies (Zimmerman, 2009; Robinson & Segrott, 2002; Vecchio, 2016) have suggested they also want to contribute to their well-being by assuming a productive role in their society, demonstrating the potential and importance of skills development for refugees and asylum seekers. Especially for refugee and asylum-seeking mothers, who are the key factors in children's early social and emotional development, increasing their capacity to be employed in green jobs and using available resources to contribute to sustainability can be expected to improve their lives and their children's.

In Hong Kong, adult RAS access to formal education is very limited. Therefore, non-formal learning organized by NGOs is the main pathway for skills development, including those relevant to SUD. However, NGOs' capacity regarding what they can offer in terms of green skills development is very low.

To fully realize greening and inclusivity, (re-) training should be accessible to all, including those from vulnerable and disadvantaged backgrounds, and as early as childhood (Strietska-Ilina et al., 2011). Training may be allowed, but the impact may not be long-lasting. Therefore, it is also important to consider an appropriate teaching and learning strategy and pedagogy to ensure impact.

3.4 Teaching Green Skills

Studies argue that traditional pedagogy does not work for teaching sustainability. In traditional education settings, learners are passive recipients of knowledge. This knowledge is passed to them by an authority, the teacher. In such classrooms, the lectures are content- and teacher-centered and do not consider the complexity and connectedness of sustainability issues or lead students to explore real-world problems. It leads to a gap between knowledge and its application. This is a similar problem that Pavlova and Chen (2019) observed in TVET classrooms in Hong Kong. Therefore, an approach and strategy for developing a vision that cares, caters to, and transforms a healthy environment for all is necessary. Many authors argue that the best approach to teaching and learning sustainability should be 'transformative' (Pavlova, 2012; Tilbury, 2011; Maclean & Pavlova, 2017).

The transformative approach has been found effective in various education areas besides sustainability, such as peace education, health education, and environmental education (Tilbury, 2011). The goal of transformation is self-actualization or personal transformation, which can be achieved through developing consciousness about socio-ecological processes and issues. Pavlova (2012) calls it social emancipatory transformation, and Vernadsky (1945) considers it a global consciousness (or as he calls 'noosphere wisdom'). A transformative approach is regarded as the most appropriate way of teaching and learning about sustainability for the reasons below.

(i) It advocates for *critical reflection* so the learner can identify their reality and agency to transform society.
(ii) It has a *liberational approach to teaching* because the problem is cognitively identified, and discussions arise.
(iii) It considers teacher-student relationships on an equal and horizontal level (Pavlova, 2012).

Furthermore, the three strategic objectives defined for Refugee Education 2030 necessitate a similar approach. The goals for refugee education 2030 (UNHCR, 2019a, 2019b) are:

1. Promote equitable and sustainable inclusion in national education systems for refugees, asylum seekers, returnees, stateless, and internally displaced persons.
2. Foster safe, enabling environments that support learning for all students, regardless of legal status, gender, or disability.
3. Enable learners to use their education toward sustainable futures (UNHCR, 2019a, 2019b).

Of these objectives, the third one is significantly related to the theme of this book and the chapter. The sub-goals of this objective state that as a result of education, children and youth are expected to serve as national champions for their communities and play an active role in advocating for and developing solutions that contribute to sustainable futures nationally and globally. Another expected result targeted explicitly for women and girls is to pursue and complete tertiary education on par with men at all levels, including TVET, and engage in labor market opportunities and apprenticeships, including in technical, professional, and other fields not traditionally open to women, and exercise ownership over management and control of household resources and cash-based interventions (CBI) (Please see UNHCR, 2019a, 2019b, Appendix A for a complete list of expected results).

Pavlova's (2012) ESD-informed, social emancipatory transformation pedagogy attempts to establish a capacity for restructuring the distorted, profit-indexed approach with a healthier and value-centered approach towards human beings and nature in all domains of teaching and learning. For this pedagogy, multiple stakeholders have a role in solving SD issues, including refugee-related and environmental stakeholders. In this respect,

> ESD is far more than teaching knowledge and principles related to sustainability. ESD, in its broadest sense, is education for social transformation with the goal of creating more sustainable societies. ESD touches every aspect of education including planning, policy development, programme implementation, finance, curricula, teaching, learning, assessment, administration. ESD aims to provide a coherent interaction between education, public awareness, and training with a view to creating a more sustainable future (Shaping the Education of Tomorrow, 2012).

For this reason, Education for Sustainable Development (ESD) Pedagogy is suggested as an effective way of teaching and learning about generic green skills and sustainability (Down & Ferguson, 2021).

3.4.1 Education for Sustainable Development (ESD) Pedagogy

Freire argues that the aim of knowledge transfer, where only one person is in control, such as the teacher, is to control the population and maintain the current system so that the elite can retain their power (Freire, 1970). This approach is considered oppressive because it presumes students do not know anything. However, the source of knowledge cannot be confined to a particular group or power only. It is accepted worldwide that indigenous knowledge systems provide us with traditional knowledge, observations, practices, and attitudes toward living in harmony with nature and are considered a key to successful development (Warren & Cashman, 1988; Priyadarshini & Abhilash, 2019; UNDP, 2024). For this reason, teaching and learning about sustainability, whether it is through formal environmental education or non-formal and informal education for sustainable development, requires a considerate, inclusive, and transformative approach where knowledge is constructed through a dialogue of the stakeholders (in his book is the oppressed and the oppressor). Education should not be decided, transferred, and mediated by 'authorities to 'the public,' or by 'authorities about 'the public.' Rather, it should be practiced by 'authorities' with 'public.' This is the liberating aspect of education because it allows (the oppressed) to 'create, construct, wonder and venture' (Fromm, pp. 52–53, cited in Freire, 1970). Therefore, when designing a response to climate change-related issues, communicating with the public about their problems and providing them an option to contribute to the solution is a practice of freedom. In the case of refugees and asylum seekers, whose opinions and experiences are often unheard of, it can be considered a form of empowerment as they are given a choice to take part in such dialogue.

However, the motivation to participate depends on an existential, concrete, and present situation. In the case of environmental issues, the problem is existential and concrete as people are forced to migrate due to floods, earthquakes, and other natural disasters. The problems stemming from environmental issues may not always be visual, felt directly, or considered in association with other issues. For this reason, an intellectual understanding of the situation and a reflection on values and behavior are essential to the solution. Freire (1970) refers to action and reflection as equal dimensions of the liberating praxis of education. In his work, action without reflection is considered an action just for the sake of acting; reflection without action is considered verbalism, where the word is an empty "blah." (See Chap. 3 of Freire (1970) for further information).

Caliskan (2020) argues that nature can heal itself and find a balance because other elements (such as animals, plants, mines, etc., but not human beings) in the ecosystem perform tasks assigned to them and do not harm the ecological balance for 'economic profit.' However, competition and conflict among societies and communities, which often stem from power relations and profit orientation, harm the environment. The ambition of those with power leads to economic problems, oppression, wars, anarchy, and terrorist incidents. Such economic or socio-political issues can initiate environmental problems like air pollution, water pollution, and resource use, resulting in increased refugees and asylum seekers. Environmental and refugee-related problems are two of the most critical issues visible worldwide and influence everyone's lives. Therefore, this pedagogy emphasizes that issues are intricate and connected like chainrings. Caliskan (2020) also adds that humans can restore the damage they have done to the environment. Therefore, for a successful and transformative experience, a revolutionary leader (in our case, a teacher) combined with co-intentional practice is necessary. In this approach, the oppressed (students/learners) and the oppressor (teacher/leader) come to know critically how a specific knowledge comes to being, share each other's perspectives, and re-create the reality through joint reflection and action. That way, the oppressed (in our case, RAS learners) become committed to the cause (Freire, 1970). So, this pedagogy's empowering/liberating aspect is that human beings should realize how they can restore balance through their actions after reflection and discussion. That way, development can be sustained.

Similarly, Gamage (2022) argues that this type of approach treats the environment as a social construct and deems it an important stance to invite people to act because if it is viewed as such, then it is accepted that certain parts of it can be transformed or altered depending on whichever social construct is on operation (i.e. individual, collective, state-led, multi-nationally led) (Gamage, 2022, p. 34). Hence, the author suggests that education should equip learners with a critical lens and literacy to look at what is happening, whose interests are being served, where we are going, and where we are standing among all this so that they can participate in the resolution.

Labadi (2022) suggests considering culture, nature, and the heritage of conservation and management from a holistic perspective. It is argued that local communities have their ways of preserving nature. However, when an ivory carver is asked to stop his work because it is not serving sustainability and not offered an alternative way, this does not foster a supportive attitude or behavior toward the cause in him. Therefore, the author argues that people/local communities must see the benefit of environmental protection for themselves. The interpretation of this in an educational context can be sharing the realities of local communities and co-creating a solution that respects their livelihood and the environment. This brings us to the other aspect of ESD learning: entrepreneurship. Developing entrepreneurial skills is part of generic green skills and a necessary driver for inclusive and green growth (OECD, 2012). Thus, entrepreneurship development should be one of the focuses for supporting the greening of economies, as well as teaching and learning programs about generic green skills.

Stevenson (2006) suggests that to transform society and the environment, we must create a space where educators are involved in discourse. It should be informed historically, pedagogically, and politically by practice in diverse cultural contexts because we cannot address a vision of a sustainable society without simultaneously considering the present environmental issues that need to be tackled. To achieve this praxis, we need to adopt a pedagogy that communicates with learners and enables them to reflect on and act on this problem. Table 3.2 indicates ESD's proposed transformation in teaching and learning.

In support of the above shifts in teaching and learning approaches for ESD, Tikly (2019) suggests we need to understand what is unsustainable if we are to identify a vision for sustainability. Stevenson (2006) argues that an intensive, in-depth study of fewer, less complex, local environmental problems or issues can develop more critical and enduring outcomes than simply transmitting discrete environmental facts and concepts. Such an approach would make it easy for students to connect topics to their lives, increase their critical thinking capacity, and help them improve their concern for environmental issues (Stevenson, 2006).

On the other hand, Bonnett (2002) adopts a metaphysical approach to ESD. He interprets sustainability as a frame of mind, which means being aware ("conscious"

Table 3.2 Educational shifts proposed by ESD

FROM	TO
Passing on knowledge	Understanding and getting to the root of issues
Teaching attitudes and values	Encouraging values clarification
Seeing people as the problem	Seeing people as facilitators of change
Sending messages	Dialogue, negotiation, and action
Behaving as the expert—Formal and authoritarian	Acting as a partner—Informal and egalitarian
Raising awareness	Changing mental models that influence decisions and actions
Changing behavior	More focus on structural and institutional change

Source: Tilbury, 2011

in his words) of how one fits in one's surroundings (he would possibly use the terms "cosmos" or "eco-system" instead of surroundings) and responds to it critically (Bonnet calls this process 'flourishing'). Therefore, he argues that the goal of education for sustainability should foster responsive and responsible, practical, flourishing whereby attitudes and values are expressed in life (outside school borders). However, as Hungerford and Volk (1990) argue, behavior change as a result of education requires a new model of instruction where environmental "issues[2]" are at the heart of the instruction. Students are given the opportunity of "ownership" and "empowerment," so they are fully invested in an environmental sense and become responsible, active citizens. Therefore, to achieve learning and teaching objectives, it is imperative that all people, including refugees and asylum seekers, are valued and included in the discourse.

In refugee contexts, especially for adult refugees, four competencies are of key importance: language, occupation, personal development, and integration. The first two are usually acquired in non-formal (courses, training, etc.) or formal education (if certification is required); it is the latter that refugees have difficult access to, just like in the case of the labor market. Personal development and integration usually happen spontaneously, as these competencies are more difficult to define, and they develop systematically. It is argued that a well-prepared mentor and a well-thought process focusing on leaner development covering specific themes may be effective when it comes to supporting self-development and mitigating cultural shock consequences (Jarymowicz, 2023). In this respect, social entrepreneurship is particularly relevant to refugees and asylum seekers as entrepreneurial actions may help them achieve economic and social stability and integrate into their new environment.

Williams (2001) looks at the quality of refugees' education. Although the points were mainly for children and formal learning, some of his analysis's points and scope were relevant to our context. For instance, he argues that refugees' readiness for learning is likely to have been compromised by their traumas. In the case of RAS in HK, the stress of living with the minimum they receive leads to traumatic experiences (see Chap. 2 for RAS in HK). He further argues that school quality can be understood in several ways and involves four interacting sets of factors: the characteristics of the child/learner (individual differences such as age, gender, nutritional status, etc.); supporting inputs (system-level factors such as the language of instruction, parental and community support, teacher qualifications, etc.), enabling conditions (such as the effective school leadership, an organized curriculum, maximized learning time in schools, incentives for academic success), and the teaching-learning process (such as use of various techniques, homework, students assessments etc.) (Williams, 2001). Although these are defined for formal schools, they influence non-formal learning settings. For instance, instead of school leadership, NGO

[2] They note that environmental issues and environmental problems differ in meaning. While the problem indicates a risk, i.e., endangered species, issues highlight the difference in values and beliefs among humans about what should be done concerning the problem.

leadership can be considered as in the case of Hong Kong; it is the NGOs providing such education to the concerned group. Therefore, quality education and training must attend to all these dimensions regardless of the setting.

3.4.2 Implementation of ESD Pedagogy

In implementing such pedagogy, various teaching and learning activities can be considered. Since ESD has been taught, more learning styles and combinations have evolved (Shaping the Education of Tomorrow, 2012). The global monitoring and evaluation survey (GMES) identified nine types or forms of learning associated with ESD. Some can be considered conventional (e.g., transmissive learning, disciplinary learning), and some more cutting-edge (e.g., multi-stakeholder social learning and systems thinking-based learning). The nine forms are described briefly below:

> ***Discovery learning*** involves immersing learners in a rich context where they encounter some element of mystery. They become curious and begin to make sense of their experience through exploration.
>
> ***Transmissive learning*** uses didactic skills (e.g., presenting, lecturing, storytelling) and supporting materials (e.g., workbooks, instruction forms, visuals) to transfer a body of knowledge, set of rules, or code of conduct to learners.
>
> ***Participatory/collaborative learning***—Although participation and collaboration are not identical, both emphasize working together with others and being active, not passive, participating in the learning process and focusing on resolving a joint issue or task.
>
> ***Problem-based learning*** is focused on solving actual or simulated problems to understand the issue better or find ways to make real-life improvements. Issues are either identified by the learners or pre-determined (by teachers, experts, and commissioning bodies).
>
> ***Disciplinary learning*** takes questions of a disciplinary nature (e.g., geographical or biological) as a starting point to understand underlying principles better and expand the knowledge base of that discipline.
>
> ***Interdisciplinary learning*** takes issues or problems as a starting point and then explores them from different disciplinary angles to arrive at an integrative perspective on possible solutions or improvements.
>
> ***Multi-stakeholder social learning*** brings together people with different backgrounds, values, perspectives, knowledge, and experiences from inside and outside the group to initiate the learning process and set out on a creative quest to solve problems without ready-made solutions.
>
> ***Critical thinking-based learning*** exposes the assumptions and the values people, organizations, and communities live by and challenge their merit from a normative point of view (e.g., animal well-being, eco-centrism, human dignity, sustainability) to encourage reflection, debate, and rethinking.
>
> ***Systems thinking-based learning*** looks for connections, relationships, and interdependencies to see the whole system and recognize it as more than the sum of its parts, as well as to understand that intervention in one part affects other parts and the entire system (Shaping the Education of Tomorrow, 2012, p 25-26).

Based on the learning methods mentioned above, discovery learning, systems thinking-based learning, critical thinking-based learning, interdisciplinary learning, problem-based learning, and participatory/collaborative learning were the most favored by the Global Monitoring and Evaluation Survey (GMES) respondents. In terms of teaching, Maclean and Pavlova (2017) suggest that teachers should provide an opportunity to engage learners in pro-environmental activities, and they emphasize the importance of values that are not related to economic imperatives, such as valuing the other person, moral responsibility, and establishing a non-mechanistic relationship with nature.

Using Bloom's taxonomy for educational objectives, Sern et al. (2018) observe (generic) that green skills cover all three domains of learning, which are necessary for learners (in their article, they are 'workers') to promote sustainable development in the social, economic, and environmental domains (Sern, Zaime, & Lee, 2018). For example, in a classroom where generic green skills are taught, Bloom's taxonomy and each domain can be covered in the following way: to tap into cognitive development when learning about waste management, issues about how it applies to daily life and industry, the challenges faced, and responses to these problems are covered. These also reflect the cognitive competencies of generic green skills. In terms of the affective domain, realizing one's disposition towards the environment and waste production (such as eco-centric or anthropo-centric orientation) and developing a sense of belonging to nature, as well as communicating effectively in conflicting issues that arise from this issue, can be fostered. These fall under interpersonal competencies in generic green skills. Regarding the psycho-motor domain, developing technical skills is necessary to produce renewable energy sources, transfer and top up current skills in a new job, and turn waste into a green-friendly opportunity/action such as in social entrepreneurship, as covered under intrapersonal competencies in generic green skills. Sen (1999) argues that the sustainability of development depends on people's role as agents of change. Furthermore, the role of education systems is to develop the capabilities of current and future generations to achieve sustainable livelihoods and well-being within peaceful and democratic societies. This requires considering disadvantaged groups' institutional barriers to realizing valued capabilities and functions (Tikly, 2019).

Educational institutions have traditionally been far more focused on effective individual development and a particular way of teaching (that stresses the transmission of knowledge and teacher-centered side) than fostering collaborative skills and the collective agency needed by communities and social movements. So, asking teachers to change how and what they teach should change the organizational and political context (Stevenson, 2007). Moreover, even if it is politically achieved, negotiating and enacting change in practice is much more complicated than negotiating change in policy and rhetoric because the former challenges vested interests in the status quo. In addition, the amount of space devoted to participation, self-determination, and independent thinking has implications for how ESD is perceived and realized. It is reported that when this space is narrow, teaching and learning practices tend to be more transmissive and teacher centered. Although this approach may be advantageous in reaching more people and achieving a desirable result, such as behavioral change about sustainability, it does not promote a meaningful engagement/internalization with the subject. However, ESD tends to prevail when the space for participation is broad, more interactive, student-centered, and transformative. It has been argued that this mode of teaching and learning leads to capacity building, empowerment, and behavior change (Shaping the Education of Tomorrow, 2012). Therefore, all human beings can be agents of change, including RAS. However, to be a facilitator of change, one needs to consider oneself as the partner to the solutions concerning environmental problems according to ESD thinking (Tilbury, 2011). This consideration of responsibility and the necessity for a collective effort is relevant to the transformative pedagogy that Freire (1970) posits. Both ESD and transformative pedagogy require reflection about the root cause of the problems and negotiations with society to (re-)define the established values and actions. A number of techniques can be used. To cultivate problem-solving, a common activity is to use the '5 WHYs'. This technique asks the learner to ask why many times to get to the bottom of the problem. Similarly, a problem tree analysis can also help map the causes and effects of a problem.

It is also vital for teachers and learners to understand that ESD does not lead to competition between different groups of society, nor does it require one group to prove itself to be more worthy because, at its core, ESD takes an egalitarian approach. Moreover, if RAS, or any other marginalized group, were to participate in a learning activity with the feeling that they need to prove themselves to be equal to dominant groups, this could create a stress factor and diminish the learning experience (Hooks, 1994). Therefore, ESD should allow for collaborative learning, where new ideas/values are created while developing generic green skills.

However, when considered with Esser (2004)'s approach as mentioned earlier in the chapter, it is important for refugees to see the positive impact of their investment, whether it be time or money, in participating in green skills development training. Understanding that they are destitute, the training must offer them something valuable in return, such as money or feeling accomplished etc.

3.5 The Potential of Entrepreneurship for Empowerment of Refugees

In this regard, supporting refugees through entrepreneurship has proven to be a successful strategy. It does not only lead to achieving inclusive economic development but also contributes to the integration of communities. Although not all refugees are born entrepreneurs, nor should they be, research suggests that many migrants are naturally entrepreneurial because they have chosen to leave their countries of origin and build a new life abroad. They have a greater tolerance for risk and hard work. Some migrant and refugee groups may tend towards entrepreneurship because of their personal, family, or cultural backgrounds. A study that looked at entrepreneurial activities in refugee camps suggested that half of the respondents had entrepreneurial experience before, and more than 80% of respondents showed a willingness to participate in micro-entrepreneurship, highlighting the potential for this group (Kachkar, 2019). It has also been suggested that the entrepreneurial activity of migrants and refugees stimulates entrepreneurship among others (whether native or foreign-born) (UNCTAD, 2018).

A qualitative study from Turkiye looked at the role of entrepreneurship in facilitating the integration process of refugees and refugees' motivation to participate in entrepreneurial activities in the hospitality industry and found that entrepreneurial activities help refugees integrate with the socio-economic fabric of the host country. The study also reveals that refugees' motivation varied from survivability, desire for independence, and availability of resources (Shneikat and Alrawadieh, 2019). Another example from Ecuador demonstrates that creating small businesses can help build capacity and increase resilience by creating employment and a feeling of contribution on the community and personal level (Sánchez Piñeiro & Saavedra, 2016; Fitzsimons, 2011). They can benefit the country's economy and create jobs at the national level.

From a gender-sensitive lens, labor participation among refugee women is reportedly low in the Netherlands due to needing to take care of household duties or holding traditional values about gender roles—an example of an informal institution (Bakker et al., 2017). Similar to the Netherlands, a study from Uganda showed that the number of women entrepreneurs is negligible. Although women do not necessarily own a shop, they contribute to their home economies in less visible ways, such as tailoring, candy making, and other handcrafts (Betts, Bloom, & Weaver, 2014). Therefore, women may have choices, but their choices to attain greater power in life seem lower than those of men. However, this does not mean they are not good businesspeople.

A program for refugees in Australia revealed that only 20% of participants were women. However, women's businesses were more profitable than men's (Collins, 2017).

Therefore, although country-specific situations can differ, women's skills and potential can be improved through entrepreneurial skills training, and their contribution to the economy can be made more visible. For example, the sectors' need led by greening is not met in many developing countries due to a lack of training, skills, and labor (Strietksa-Ilina, Hoffman, Haro, & Jeon, 2011). So, there is potential for women refugees to be trained in entrepreneurship and see how they can contribute to greening. Such a positive contribution may strengthen refugees' resilience and the communities hosting them (SDG Knowledge Platform, n.d.).

However, it can be argued whether focusing on empowerment could overlook this community's more pressing need for immediate support and put an additional burden on the adjustments of refugees. This could be the case, especially when people run for their lives in a war conflict. However, this is not the case for many RAS in HK. If the organization of training and employment is considered a result of shared responsibility between citizens, governments, and refugees, it can benefit RAS and allow them to contribute to the city. Chell (2007) argues that "entrepreneurialism is about the social construction of reality through the envisioning of possible futures that are both social and economic". Often, ideologically rooted self-enterprise approaches take little account of insecure labor markets or refugees' capabilities to transform humanitarian assistance and livelihood opportunities into something sustainable and meaningful for them (Field et al., 2017). Refugees' variable urban experiences call for a more flexible and longer-term framing of self-reliance and community resilience. The goals need to be less about individual entrepreneurial self-sufficiency and more about refugees' collective and collaborative ability to convert places, services, and opportunities into human well-being (Field et al., 2017), particularly when societies and cities are concerned about sustainability. Therefore, collaboration between different societal stakeholders is required to cultivate positive communication and demonstrate an empowering approach (World Economic Forum, 2017) for the well-being of all communities.

As illustrated by the above examples, there is generally a lack of environmental orientation in RAS's entrepreneurial training or programs. However, several successful examples related to social entrepreneurship consider the environmental impacts. A study by Nayak, Salovaara, and Wade (2019) investigated entrepreneurship among Tibetan refugees in India and introduced social entrepreneurship as a concept that was part of their refugee entrepreneurship program. Some participants reported interest and would like to learn more about this concept (Nayak, Salovaara, & Wade, 2019). Therefore, below, we explain what it is and how it can correspond to SUD.

3.5.1 What Is Social Entrepreneurship?

The economies in most countries comprise three sectors: the public sector (a.k.a. government), the private sector (a.k.a. business), and the nonprofit sector (a.k.a. civil society). But now, a fourth sector of the economy is emerging, combining market-based approaches of the private sector with the social and environmental aims of the public and non-profit sectors to address pressing issues. Endeavors in this sector, also known as for-benefit enterprises, come in various models, from mission-driven businesses, social enterprises (the focus of the below section), and sustainable businesses, among many others. As the world faces more significant challenges every day—from increasing gaps in inequality to pandemic health threats and a global climate crisis—growing this emerging sector of the economy focusing on solving the problems of people and the planet is gaining traction.

The mission of a social enterprise, as opposed to a conventional one, is to create social and environmental change while making a profit. Lately, this sector has been considered under the fourth sector (for benefits). Therefore, it was chosen as a tool to empower RAS in the study presented in the book. Social enterprises focus on society's benefits and the desire to create change. In social entrepreneurship, social value and social change are somewhat subjective because what people value or at what level they want to make a change can be in conflict (Singh, 2016). Since the main target is to meet social needs and create social value, the context in which entrepreneurs seize such opportunities can reverberate in various sectors, such as the environment, health, and education. Social entrepreneurs also utilize financial, personal, and other resources, and they are not limited by what is available at hand because they can creatively mobilize resources. The main difference between conventional and social entrepreneurs is that the latter tend to mobilize resources for building capacity and solving problems related to resourcefulness. Social entrepreneurs mobilize different resources, such as social and cultural resources.[3] (e.g., founding team members, paid employees, and volunteers), financial resources, and other resources (mainly land and infrastructure) to start their social enterprise operations (Fig. 3.1).

The World Bank has estimated that tens of trillions of private finance and investment dollars need to flow toward global development goals by 2030. This is referred to as the "billions-to-trillions gap". Expansion of the fourth sector is, in fact, what is required to close this gap because fourth sector development (vs. traditional development) advances scalable, market-based solutions across all the SDGs while at the same time reducing the negative social and environmental externalities associated with growth. Also, the fourth sector addresses the root causes underlying each of the SDGs, which are often related to negative externalities associated with the current approach to development that has become the global norm—by embedding sustainability and human rights into the DNA of businesses, thereby yielding more sustainable, inclusive, ethical development outcomes as they grow. And finally, given the breadth and rapid growth potential of the fourth sector—there are tens of millions of social-purpose enterprises starting up around the world each year—developing the fourth sector can unleash entrepreneurship and innovation at a massive scale to address challenges across the full spectrum of SDGs globally.

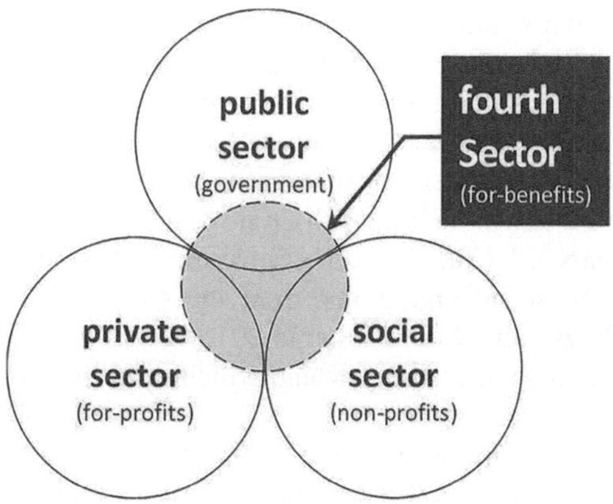

Fig. 3.1 Social enterprises, the fourth sector. Source: Fourth Sector Group, 2014

[3] In the original text, Singh (2016) refers to these as human resources. However, for consistency purposes related to the resources we name in this book, we consider human resources to be social and cultural as they can be formed via social and/or ethnic networks.

However, as identified by Singh (2016), SE entrepreneurs often have limited access to resources at the initial stages of their journey. Due to constraints, especially regarding access to financial resources, people start with a minimum budget and few employees (Singh, 2016). However, as Bradley et al. (2012), capital is not the problem for social enterprises, but the "idea problem" is. However, innovative and new ideas are central to economic change (Bradley et al., 2012). Katre and Salipante (2012) support this notion, and their study shows that successful social enterprises have a solid commitment to their mission and persistently refine economic and social opportunities by seeking extensive feedback and diversifying their network to make the business work.

Social entrepreneurship can contribute to developing and implementing green economies, provided the agent/entrepreneur is committed to it. Entrepreneurship can provide an opportunity for refugees to contribute to greening creatively, although their access to resources is limited. The success of such a social mission depends on one's knowledge and specific skills and the ability to learn how to negotiate in changing or unfamiliar circumstances, manage risks, and learn from these experiences (Singh, 2016). Grom & Chahine argue that developing refugees' and asylum seekers' social entrepreneurship for establishing or maintaining a social enterprise can help them maintain their daily lives (Grom & Chahine, 2017).

3.5.2 Social Enterprises in Hong Kong

Although the logic behind social enterprises is to make an environmental and social impact, their definition, operational models, and legal forms may differ from one country to another. According to the government in HK, a social enterprise is defined as "a business to achieve specific social objectives such as providing the services (e.g., support service for the elderly) or products needed by the community, creating employment and training opportunities for the socially disadvantaged, protecting the environment, or funding its other social services through the profits earned. Its profits will be principally reinvested in the business for the social objectives that it pursues" (Yu, 2016). The government has injected HK$450 million in seed money into social enterprises from the early 2000s onwards. It was reported by the Legislative Council that in 2015, there were only 78 social enterprises per million population in Hong Kong, which is far behind the figure of 1083 in the UK, for example.

In addition, in Hong Kong, local social enterprises are highly concentrated in lower value-added activities, such as catering (30%), medical care (26%), and retail and personal services (25%). The lack of SE development resulted from a 'lack of expertise in running a business, 'limited government support and guidance in the application stage,' 'limited mentorship support,' and 'lack of public service contracts' (Yu, 2016). Considering Grom & Chahine's (2017) argument that SEs can help RAS maintain their lives and the lack of environmental services concentration in Hong Kong's social enterprises, the provision of skills training to RAS to learn about generic green skills and as part of it, social entrepreneurship, can be useful to enable them to contribute to city's effort for sustainability. This would also empower them as it would provide them with a chance to join or not.

3.6 Summary

Education, capacity building, and skills development have been stressed repeatedly in discourses on the greening of economies, SDGs, and inclusive SUD, as indicated above. The green transition suggests that changes in the economic trajectory will influence the nature of jobs (International Labour Organization, 2011). There will inevitably be significant changes in labor demand and the skills required for new and existing positions (Cedefop, 2014); (Strietksa-Ilina, Hoffman, Haro, & Jeon, 2011). International and regional cooperation agencies, such as ILO, OECD, and CEDEFOP, have focused on developing labor force skills, knowledge, and competencies to respond to product, service, and process changes. These are critical for capturing the green transition's social, financial, and environmental benefits (Cedefop, 2014). For this reason, countries should invest in understanding the employment potential of the green transformation and the corresponding needs in education and training and then steer skills development systems towards meeting those new needs in a way that caters to all groups in society (International Labour Organization, 2011), without neglecting marginalized groups.

Hong Kong's high economic status does not guarantee social inclusion. To address sustainability's economic, social, and environmental aspects, HKSAR has implemented some initiatives and faces some constraints. Therefore, in addition to regulations at the state level to achieve SUD, values, skills, and actions on the individual level should also be cultivated to support this transformation. In this regard, developing generic green skills for all is the most effective. However, values and behavior cannot be developed separately. Each learning setting, namely formal, informal, and non-formal, has advantages and disadvantages. What matters is the approach and pedagogy taken in these settings.

We examined the relevant literature to help us design an intervention for our study. We were influenced by Pavlova's (2012) and Freire's (1970) discussions about educational pedagogies that are emancipatory and transformative when designing the approach for training. This helped us identify ESD as the appropriate pedagogy as it cohered with the objectives of our study. In addition, the problems related to environmental education in Hong Kong revealed that classes are teacher- or content-centered, and there is a disconnect between sustainability and real-world issues. The intervention design by Pavlova and Chen (2019) was helpful as it could be applied to our study in a non-formal setting and address the problems identified earlier.

The next chapter presents the study's findings on the concept of resources and discusses their impact on RAS well-being within the context of Hong Kong.

References

Alsop, R., Bertelsen, M. F., & Holland, J. (2006). *Empowerment in practice from analysis to implementation*. The World Bank. Retrieved from https://elibrary.worldbank.org/doi/pdf/10.1596/978-0-8213-6450-5

Arbuthnott, K. D. (2009). Education for sustainanle development beyond attitude change. *International Journal of Sustainability in Higher Education, 10*(2), 152–163.

Asian Development Bank. (2020). *Overview of gender equality and social inclusion in Nepal*. ADB. Retrieved from https://www.adb.org/sites/default/files/institutional-document/32237/cga-nep-2010.pdf

Bakker, L., Dagevos, J., & Engbersen, G. (2017). Explaining the refugee gap: A longitudinal study on labour market participation of refugees in The Netherlands. *Journal of Ethnic and Migration Studies, 43*(11), 1775–1791. Retrieved from https://www.tandfonline.com/doi/full/10.1080/1369183X.2016.1251835

Becker, G. (1995). The economic way of looking at behavior. In R. Febrero & P. Schwartz (Eds.), *The essence of Becker* (pp. 633–658). Hoover Press.

Betts, A., Bloom, L., Kaplan, J., & Omata, N. (2014). *Refugee economics: Rethinking the popular assumptions*. University of Oxford. Retrieved from https://www.rsc.ox.ac.uk/files/files-1/refugee-economies-2014.pdf

Bhowmik, M. K. (2014). Ethnic minority students' education in Hong Kong: Rhetoric, issues and priorities. *International Journal of Comperative Education and Development, 16*(2), 126–140.

Bloom, B. S., Engelhart, M. D., Furst, E. J., Hill, W. H., & Krathwohl, D. R. (1956). *Taxonomy of educational objectives, handbook I: The cognitive domain*. David McKay, Inc.

Bonnett, M. (2002). Education for sustainability as a frame of mind. *Environmental Education Research, 8*(1), 9–20. https://doi.org/10.1080/13504620600942683

Bradley, S. W., McMullen, J. S., Artz, K., & Simiyu, E. M. (2012, June). Capital is not enough: Innovation in developing economies. *Journal of Management Studies, 49*(4), 684–717. https://doi.org/10.1111/j.1467-6486.2012.01043.x

Brown, O. (2008). *Migration and climate change*. International Organization for Migration.

Brundtland Commission. (1987). *Report of the world commission on environment and development: Our common future*. World Commission on Environment and Development. Retrieved from https://sustainabledevelopment.un.org/content/documents/5987our-common-future.pdf

Caliskan, N. (2020). *Kur'an'da Cevre Bilinci*. Arastirma Yayinlari.

CEDEFOP. (2010, July). *Briefing note*. Retrieved from skills for green jobs: Chrome-extension://efaidnbmnnnibpcajpcglclefindmkaj/https://www.cedefop.europa.eu/files/9024_en.pdf

Cedefop. (2014). *OECD Greener skills and jobs*. OECD Publishing. Retrieved from https://read.oecd-ilibrary.org/industry-and-services/greener-skills-and-jobs_9789264208704-en#page4

Chan, K., & Siu, Y. P. (2015). Urban governance and social sustainability: Effects of urban renewal policies in Hong Kong and Macao. *Asian Education and Development Studies, 4*, 330–342.

Chazan, B. (2003). The philosophy of informal Jewish education. *The encyclopedia of pedagogy and informal education*. Retrieved from https://infed.org/dir/the-philosophy-of-informaljewish-education/

Chell, E. (2007). Social enterprise and entrepreneurship: Towards a convergent theory of the entrepreneurial process. *International Small Business Journal, 25*(1), 5–26. https://doi.org/10.1177/0266242607071779

Chisholm, L. (2005). The making of South Africa's National Curriculum Statement. *Journal of Curriculum Studies, 37*(2), 193–208. https://doi.org/10.1080/0022027042000236163

Choi, M. (2024, February 13). *Green jobs: ESG and sustainability skills in high demand as employers struggle to find qualified candidates*. Retrieved from SCMP: https://www.scmp.com/business/companies/article/3251763/green-jobs-esg-and-sustainability-skills-high-demand-employers-struggle-find-qualified-candidates

Collins, J. (2017). *Ignite: From refugee to entrepreneur in Sydney final evaluation report on the SSI ignite small business start-ups program*. Settlement Services International & University of Technology Sydney.

Down, L., & Ferguson, T. (2021). *Education for sustainable development in the Caribbean: Pedagogy, processes and practices*. The University of the West Indies Press. Retrieved from https://viewer-ebscohost-com.eproxy.vtclib9.vtc.edu.hk/EbscoViewerService/ebook?an=3573698&callbackUrl=https%3a%2f%2fresearch.ebsco.com&db=e000xww&format=EK&profId=ehost&lpid=&ppid=&lang=en&location=https%3a%2f%2fresearch-ebsco-com.eproxy.vtclib9.vtc.edu

Education Bureau. (2016). *Environmental report 2016*. Retrieved from https://www.edb.gov.hk/attachment/en/about-edb/publications-stat/environmental-report/Environmental%20Report%202016_%20eng_fair.pdf

Education Bureau. (2010, October 14). *Current work of EDB on environmental education*. Retrieved from education for sustainable development in Hong Kong School_Paper no. 08/10: https://www.enb.gov.hk/sites/default/files/susdev/html/en/council/EPSCPaper08-10e.pdf

Environmental Protection Department. (2024). *Pilot programme on smart recycling systems*. Hong Kong Waste Reduction Website. Retrieved from https://www.wastereduction.gov.hk/en-hk/pilot-programme-smart-recycling-systems

Esser, H. (2004). Does the "new" immigration require a "new" theory of intergenerational integration? *International Migration Review, 38*(3), 1126–1159. Retrieved from https://www.jstor.org/stable/pdf/27645428.pdf?refreqid=excelsior%3A9a1fea9cf31bf5811e8f1197fd336eb9

Field, J., Tiwari, A. D., & Mookherjee, Y. (2017). *Urban refugees in Delhi: Self-reliance can't be exclusively entrepreneurial*. IIED Library. Retrieved from https://pubs.iied.org/17427iied

Fitzsimons, A. (2011). What is empowerment? In A. Fitzsimons, M. Hope, C. Cooper, & K. Russell (Eds.), *Empowerment and participation in youth work*. Exeter: Learning Matters Ltd.

Fourth Sector Group. (2014). *What are for-benefit organizations?* Retrieved from Fourth Sector Group: https://www.fourthsector.org/for-benefit-enterprise

Freire, P. (1970). *Pedagogy of the oppressed*. (M. B. Ramos, Trans.). Penguin Books.

Gamage, S. (2022). Environmental education for sustainability: Environmental literacy and action-oriented teaching. In M. K. Verma (Ed.), *Environment and sustainable development: Perspectives and issues* (pp. 31–49). Routledge.

GIZ. (2019). *Life skills and training for refugees and host communities*. Retrieved from GIZ: https://www.giz.de/en/worldwide/60103.html

Grom, E., & Chahine, T. (2017, September 27). *Applying social enterprise to refugee settings*. Retrieved from SSIR https://ssir.org/articles/entry/applying_social_enterprise_to_refugee_settings

HK Census and Statistics Department. (2023, December 21). *Hong Kong innovation activities statistics, 2022 edition*. Retrieved November 11, 2024, from innovation and technology: https://www.censtatd.gov.hk/en/wbr.html?ecode=B11100102022AN22&scode=580

Ho, D., Yau, Y., Law, C.-k., Poon, S., Yip, H., & Liusman, E. (2012). Social sustainability in urban renewal: An assessment of community aspirations. *Urbani Izziv, 23*, 125–139. https://doi.org/10.5379/urbani-izziv-en-2012-23-01-005

Hooks, B. (1994). *Teaching to transgress: Education as the practice of freedom*. Routledge.

Huckstep, S., & Kenny, C. (2022, November 4). *Solving the global green skills crisis requires a new mobility paradigm*. Retrieved from Centre For Global Development: https://www.cgdev.org/blog/solving-global-green-skills-crisis-requires-new-mobility-paradigm

Hungerford, H. R., & Volk, T. L. (1990). Changing learner behavior through environmental education. *The Journal of Environmental Education, 21*(3), 8–21. https://doi.org/10.1080/00958964.1990.10753743

International Labour Organization. (2011). *Skills for employment: Policy brief*. ILO. Retrieved from https://www.ilo.org/wcmsp5/groups/public/%2D%2D-ed_emp/%2D%2D-ifp_skills/documents/publication/wcms_164630.pdf

International Labour Organization. (2013). *Meeting skill needs for green jobs: Policy recommendations*. Retrieved from https://www.ilo.org/wcmsp5/groups/public/%2D%2D-ed_emp/%2D%2D-ifp_skills/documents/publication/wcms_234463.pdf

IRENA. (2022). *World energy transitions outlook 2022: 1.5°C pathway*. International Renewable Energy Agency.

Jabbar, S. A., & Zaza, H. I. (2016). Evaluating a vocational training programme for women refugees at the Zaatari camp in Jordan: Women empowerment: A journey and not an output. *International Journal of Adolescence and Youth, 21*(3), 304–3019. https://doi.org/10.1080/02673843.2015.1077716

Jarymowicz, K. (2023, March 14). *Adult refugee education. Retrieved from EPALE*: https://epale.ec.europa.eu/en/blog/adult-refugee-education#:~:text=There%20are%20four%20areas%20of%20competencies%20which%20are,to%20refugees%3A%20language%2C%20occupation%2C%20personal%20development%2C%20and%20integration

Johnson, M., & Majewska, D. (2022). *Formal, non-formal, and informal learning: What are they, and how can we research them? Research Report*. In ERIC. Cambridge University Press & Assessment. Retrieved from https://eric.ed.gov/?id=ED626005

Kachkar, O. A. (2019). Refugee entrepreneurship: Empirical quantitative evidence on microenterprises in refugee camps in Turkey. *Journal of Immigrant & Refugee Studies*, 333–352. Retrieved from https://www.tandfonline.com/doi/full/10.1080/15562948.2018.1479913?src=recsys

Karim, H. (2021, July 16). *Skills training builds futures for refugees and Pakistanis alike*. Retrieved from UNHCR Pakistan. https://www.unhcr.org/pk/13259-skills-training-builds-futures-forrefugees-and-pakistanis-alike.html

Katre, A., & Salipante, P. (2012, September 1). Start–up social ventures: Blending fine–grained behaviors from two institutions for entrepreneurial success. *Entrepreneurship Theory and Practice, 36*(5), 967–994. Retrieved from https://journals.sagepub.com/doi/10.1111/j.1540-6520.2012.00536.x

Kennedy, K., Hue, M. T., Bhowmik, M., & Ng, H. Y. (2017, June). Intersectionality and identity in an Asian context: The case of Pakistani youth in Hong Kong. *Paper presented at The CiCe Association Conference 2017: Reaffirming Citizenship Education in an Uncertain World*, VIVES.

King, K., & Palmer, R. (2010). *Planning for technical and vocational skills development*. UNESCO, IIEP.

Labadi, S. (2022). *Rethinking heritage for sustainable development*. UCL Press.

Lai, O. (2024, January 10). 6 biggest environmental issues Hong Kong is facing. *Earth.Org*. https://earth.org/biggest-environmental-issues-in-hong-kong/

Lee, K.-M., & Law, K.-Y. (2016). Hong Kong Chinese orientalism: Discourse reflections on studying ethnic minorities in Hong Kong. In A. Pratt (Ed.), *Ethnic minorities: Perceptions, cultural barriers and health inequalities* (pp. 81–116). Nova Science Publishers, Inc.

Leung, R. (2019, April 15). *Hong Kong diners go through at least 65 million tissues each day – and it makes little difference if they are eating out or at home, survey finds*. Retrieved from South China Morning Post: https://www.scmp.com/news/hong-kong/article/3006265/hong-kong-diners-go-through-least-64-million-tissues-each-day-and-it

LinkedIn Economic Graph. (2023). *Global green skills report 2023*. Be part of the green transition. Retrieved from https://economicgraph.linkedin.com/research/global-green-skills-report#:~:text=This%20report%20includes%20critical%20questions%20that%20policymakers%2C%20business,transition%20into%20jobs%20that%20help%20green%20the%20planet

Ma, X. (2019, September 11). *Proposal for Hong Kong environmental education*. Retrieved from Outdoor Wildlife Learning Hong Kong: https://drive.google.com/file/d/1CYrVnsRNAD52PYyIQ4SanIEz5vcDZzoH/view

Maclean, R. (2018a). Chapter 1: Overview. In R. Maclean, S. Jagannathan, & B. Panth (Eds.), *Education and skills for inclusive growth, green jobs and the greening of economies in Asia: Case study summaries of India, Indonesia, Sri Lanka and Viet Nam* (pp. 1–11). Asian Development Bank: Springer.

Maclean, R. (2018b). Summary of main research findings: India, Indonesia, Sri Lanka, Viet Nam. In Rupert Maclean, S. Jagannathan, & B. Panth, Education and skills for inclusive growth, green jobs and the greening of economies in Asia: Case study summaries of India, Indonesia, Sri Lanka and Viet Nam (pp. 41–78). Asian Development Bank: Springer.

Maclean, R., & Pavlova, M. (2017). What type of pedagogy is required in schools and classrooms to support sustainable green growth? A case study of Hong Kong within the international context. *Education in the Asia-Pacific Region: Issues, Concerns and Prospects (EDAP), 38*, 101–119.

Mok, O. N. A. (2011). Non-formal learning: Clarification of the concept and its application in music learning. *Australian Journal of Music Education*. Access at https://eric.ed.gov/?id=EJ952003

Nayak, G., Salovaara, I. M., & Wade, J. (2019). Self-regulated learning in refugee entrepreneurship education: A university-based program for Tibetan in India. *Diaspora, Indigenous, and Minority Education, 13*(2), 81–96. https://doi.org/10.1080/15595692.2018.1557632

Nussbaum, M. C. (2000). Introduction: Feminism and international development. In M. C. Nussbaum (Ed.), *Women and human development: The capabilities approach* (pp. 1–31). Cambridge University Press.

OECD. (2012). *Green growth and developing countries: A summary for policy makers*. OECD. Retrieved from https://www.oecd.org/dac/50526354.pdf

Paryono. (2017). The importance of TVET and its contribution to sustainable development. In *Green construction and engineering education for sustainable future. 1887* (pp. 020076-1–020076-14). AIP Publishing. Retrieved from https://scholar.google.com/citations?view_op=view_citation&hl=en&user=NGaxEI8AAAAJ&citation_for_view=NGaxEI8AAAAJ:roLk4NBRz8UC

Pavlova, M. (2012). Towards using transformative education as a benchmark for clarifying differences and similarities between environmental education and education for sustainable development. *Environmental Education Research, 19*(5), 656–672. https://doi.org/10.1080/13504622.2012.736476

Pavlova, M. (2016). Regional overview: What is the government's role in greening TVET? In: *TVET@Asia*, issue 6, 1–18. Retrieved January 30, 2016, from https://www.tvet-online.asia/issue6/pavlova_tvet6.pdf (Online).

Pavlova, M. (2018). Fostering inclusive, sustainable economic growth and 'green' skills development in learning cities through partnerships. *International Review of Education: Journal of Lifelong learning, 64*(3), 339–354.

Pavlova, M. (2019, July 14). *Greening of the economy through partnerships: Issues and impacts on skills development*. (S. McGrath, M. Mulder, J. Papier, & R. Suart, Eds.) Handbook of Vocational Education and Training, 195–218. https://doi.org/10.1007/978-3-319-94532-3_116.

Pavlova, M. (2022). Unpacking greening and skills recognition in micro, small and medium enterprises. In I. M. Pavlova & M. Singh (Eds.), *Recognizing green skills through non-formal learning* (pp. 3–22). Springer. https://doi.org/10.1007/978-981-19-2072-1_1

Pavlova, M., & Askerud, P. (2023). A euro-Asian look at challenges to innovation and the greening of industries: Implications for TVET and strategic policy formulation. *Journal of Vocational Education & Training, 76*(2), 381–405. https://doi.org/10.1080/13636820.2023.2288055

Pavlova, M., & Chen, C. S. (2019, January 8). Facilitating the development of students' generic green skills in TVET: An ESD pedagogical model. *The Online Journal for Technical and Vocational Education and Training in Asia*, 12, 1–20. Retrieved from http://tvet-online.asia/issue/12/pavlova-etal/

Pearce, D. W. (1989, June). *Sustainable development: An economic perspective*. Retrieved from IIED: Gate keeper series: Chrome-extension://efaidnbmnnnibpcajpcglclefindmkaj/https://www.iied.org/sites/default/files/pdfs/migrate/8042IIED.pdf

Powdyel, T. S. (2016). Non-formal education in Bhutan: Origin, evolution, and impact. In M. J. Schuelka & T. W. Maxwell (Eds.), *Education in Bhutan: Culture, schooling, and gross National Happiness* (pp. 169–180). Springer. https://doi.org/10.1007/978-98110-1649-3_11

Priyadarshini, P., & Abhilash, P. C. (2019). Promoting tribal communities and indigenous knowledge as potential solutions for the sustainable development of India. *Environmental Development, 32*(December). https://doi.org/10.1016/j.envdev.2019.100459

Rata, E. (2019). Knowledge-rich teaching: A model of curriculum design coherence. *British Educational Research Journal, 45*(4), 681–697. https://doi.org/10.1002/berj.3520

Rikala, P., Braun, G., Järvinen, M., Stahre, J., & Hämäläinen, R. (2024). Understanding and measuring skill gaps in industry 4.0 — A review. *Technological Forecasting and Social Change, 201*, 123206. https://doi.org/10.1016/j.techfore.2024.123206

Robinson, V., & Segrott, J. (2002). *Understanding the decision-making of asylum seekers (243; Home Office research studies)*. Home Office. Retrieved from https://www.researchgate.net/profile/Jeremy-Segrott/publication/248205716_Understanding_the_Decision-Making_of_Asylum-Seekers/links/00b49537f535421143000000/Understandingthe-Decision-Making-of-Asylum-Seekers.pdf

Ryan, D., Dooley, B., & Benson, C. (2008). Theoretical perspectives on post-migration adaptation and psychological well-being among refugees: Towards a resource-based model. *Journal of Refugee Studies, 21*(1), 1–18.

Saget, C., Vogt-Schilb, A., & Luu, T. (2020). *Jobs in a net-zero emissions future in Latin America and the Caribbean*. Inter-American Development Bank and International Labour Organization. Retrieved from https://www.ilo.org/sites/default/files/wcmsp5/groups/public/@americas/@ro-lima/documents/publication/wcms_752069.pdf

Sánchez Piñeiro, O. M., & Saavedra, R. (2016). Doing business in Ecuador. *Forced Migration Review*. Retrieved from https://www.fmreview.org/solutions/sanchez-saavedra.

Sen, A. (1999). *Development as Freedom*. Oxford University Press.

Sern, L. C., Zaime, A. F., & Foong, L. M. (2018). Green skills for green industry: A review of literature. *Journal of Physics: Conference Series, 1019*(1), 012030. https://doi.org/10.1088/1742-6596/1019/1/012030

Shaping the Education of Tomorrow. (2012). *Report on the UN decade of education for sustainable development, abridged*. United Nations Educational, Scientific and Cultural Organization. Retrieved from https://sustainabledevelopment.un.org/index.php?page=view&type=400&nr=919&menu=1515

Shneikat, B., & Alrawadieh, Z. (2019). Unraveling refugee entrepreneurship and its role in integration: Empirical evidence from the hospitality industry. *The Service Industries Journal*, 741–761. Retrieved from https://www.tandfonline.com/doi/full/10.1080/02642069.2019.1571046?src=recsys

Singh, A. (2016). *The Proces of social value creation: A multiple-case study on social entrepreneurship in India*. Springer.

Simpson, E. J. (1966). *The classification of educational objectives*. Psychomotor Domain. https://eric.ed.gov/?id=ED010368

Stevenson, R. B. (2006). Tensions amd transitions in policy discourse: Recontextualizing a decontextualized EE/ESD debate. *Environmental Education Research, 12*, 277–290. https://doi.org/10.1080/13504620600799026

Stevenson, R. B. (2007). Schooling and environmental education: Contradictions in purpose and practice. *Environmental Education Research, 13*(2), 139–153. https://doi.org/10.1080/13504620701295726

Strietska-Ilina, O., Hoffman, C., Duran Haro, M., & Jeon, S. (2011). *Skills for green jobs: A global review*. International Labour Organization. https://www.ilo.org/wcmsp5/groups/public/-dgreports/-dcomm/-publ/documents/publication/wcms_159585.pdf.

Tekleselassie, T. G. (2021). *TVET system assessment in Somali and Tigray regional states, Ethiopia, with a focus on inclusiveness*. International Labour Organization.

Thorne, A. (2020). The potential for skills development programmes to empower poverty-affected and refugee women in rural Jordan. *Journal of Vocational Education & Training, 1-17*. https://doi.org/10.1080/13636820.2020.1744691

Tian, Y., Jim, C. Y., & Tao, Y. (2012). Challenges and strategies for greening the compact city of Hong Kong. *Journal of Urban Planning and Development, 138*(2), 101–109. https://doi.org/10.1061/(ASCE)UP.1943-5444.0000076

Tikly, L. (2019). Education for sustainable development in Africa: A critique of regional agendas. *Asia Pacific Education Review, 20*, 223–237. https://doi.org/10.1007/s12564-019-09600-5

Tilbury, D. (2011). *Education for sustainable development: An expert review of processes and learning*. UNESCO. Retrieved from https://unesdoc.unesco.org/ark:/48223/pf0000191442

Trier, J. (1994). *Organizations that help the world: United Nations high commissioner for refugees*. Exley Publications.

UIL. (2022). *Making lifelong learning a reality: A handbook*. Unesco Institute for Lifelong Learning. Retrieved from https://www.uil.unesco.org/sites/default/files/medias/fichiers/2022/06/UIL_Handbook_170x240_RZ220601_Online.pdf

UN Habitat. (2018). *Working for a better urban future: Annual progress report*. United Nations Human Settlements Programme.

UN Women. (2020). *Gender mainstreaming: A global strategy for achieving gender equality & the empowerment of women and girls*. UN Women. Retrieved from https://www.unwomen.org/sites/default/files/Headquarters/Attachments/Sections/Library/Publications/2020/Gender-mainstreaming-Strategy-for-achieving-gender-equality-andempowerment-of-women-girls-en.pdf

UNCTAD. (2018). *Policy guide on entrepreneurship for migrants and refugees*. United Nations. Retrieved from https://www.unhcr.org/publications/operations/5bd31fd67/policy-guideentre-preneurship-migrants-refugees.html

UNDP. (2024, July 31). *Indigenous knowledge is crucial in the fight against climate change—Here's why*. Retrieved from Climate Promise: https://climatepromise.undp.org/news-and-stories/indigenous-knowledge-crucial-fight-against-climate-change-heres-why

UN-Habitat. (2017). *New Urban Agenda. Habitat III Secretariat*. UN-Habitat. Retrieved from http://habitat3.org/wp-content/uploads/NUA-English-With-Index-1.pdf

UNHCR. (2019a). *Education 2030: A strategy for refugee education*. The UN Refugee Agency. https://www.unhcr.org/media/education-2030-strategy-refugee-education

UNHCR. (2019b). *Refugee education 2030: A strategy for refugee inclusion*. UNHCR. Retrieved from https://inee.org/sites/default/files/resources/Refugee%20Education%202030.pdf

UNICEF. (2024). *Skills for a green transition: Solutions for youth on the move*. United Nations Children's Fund (UNICEF). Retrieved from https://www.unicef.org/media/153076/file/Skills%20for%20a%20green%20transition.pdf

United Nations. (2012). *The future we want*. Retrieved from UN Sustainable Development: https://sustainabledevelopment.un.org/content/documents/733FutureWeWant.pdf

Vecchio, F. (2016). *Asylum seeking and the Global City*. Routledge.

Vernadsky, W. I. (1945). The biosphere and the noosphere. *American Scientist, 33*(1), xxii–12. http://www.jstor.org/stable/27826043

Vona, F., Marin, G., Consoli, D., Popp, D. C. (2015, April). *Green skills*. NBER Working Paper No. w21116. Available at SSRN https://ssrn.com/abstract=2599382

Warren, D. M., & Cashman, K. (1988). *Indigenous knowledge for sustainable agriculture and rural development*. International Institute for Environment and Development (IIED). Retrieved from https://www.iied.org/sites/default/files/pdfs/migrate/X101IIED.pdf

Williams, J. H. (2001). On school quality and attainment. In J. Crisp, C. Talbot, & D. B. Cipollone (Eds.), *Learning for a future: Refugee education in developing countries* (pp. 85–108). United Nations High Commissioner for Refugees. Retrieved from https://files.eric.ed.gov/fulltext/ED480207.pdf

World Economic Forum. (2017). *Migration and its impact on cities*. World Economic Forum. Retrieved from https://www.weforum.org/reports/migration-and-its-impact-on-cities

Yang, J. (2015). *Recognition, validation and accreditation of non-formal and informal learning in UNESCO member states*. UNESCO Institute for Lifelong Learning. https://unesdoc.unesco.org/ark:/48223/pf0000232656

Yu, C.-H. (2016, November 14). Legislative Council of the Hong Kong Special Administrative Region - Social enterprises in Hong Kong and the United Kingdom. *Legislative Council Secretariat*. Retrieved from https://www.legco.gov.hk/research-publications/english/essen-tials-1617ise05-social-enterprises-in-hong-kong-and-the-united-kingdom.htm

Zimmermann, S. E. (2009). Irregular secondary movements to Europe: Seeking asylum beyond refuge. *Journal of Refugee Studies, 22*(1), 74–96. https://doi.org/10.1093/JRS/FEN048

Chapter 4
Resources and the Wellbeing of Refugees: Implications for Policies

Abstract As argued in Chap. 2, well-being is subjective and closely related to one's relationship with resources. If people can do what they value with their resources, this relates to their capability and being able to and/or having the freedom to live a life they value, demonstrating achievement of well-being, according to Sen (Development as freedom, Oxford University Press, New York, 1999). In our study we used categorization of Ryan et al. (J Refug Stud 21:1–18, 2008) foe analyzing what resources refugees had: personal resources related to physical or psychological needs; material resources related to money, property, and personal possessions, including employment; social resources related to one's emotional, informational, and tangible support in personal relationships; and cultural resources related to one's skills, knowledge and beliefs within a cultural setting (Ryan et al., J Refug Stud 21:1–18, 2008).

In our research, we intended to understand the relationship between the resources that our participants, refugees and asylum seekers, access and how these resources influence their lives in Hong Kong. We collected data through interviews, surveys, participant observations, and reflection. In this chapter, we share the data that we have analyzed based on similar categorization with Ryan et al. (J Refug Stud 21:1–18, 2008); however, collected data indicate that despite being categorized under one type, such as housing (under material resources), different types of resources can influence another aspect of well-being, such as access to education (under cultural resources).

We then discuss Hong Kong's approach at the societal level to regulating refugees' access to resources and review how equality is conceptualized in Hong Kong. We conclude that Hong Kong must replace its depoliticizing strategy towards RAS to improve RAS well-being.

Keywords Resources for RAS in Hong Kong · Conceptualization of equality in Hong Kong · Depoliticization of refugees

A. T. Saral, M. Pavlova, *Education for Refugees and Sustainable Green Urban Development*, Education in the Asia-Pacific Region: Issues, Concerns and Prospects 78, https://doi.org/10.1007/978-981-95-1680-3_4

4.1 Section 1: Findings

The presentation of the findings below are categorized under material, social, personal, and cultural resources. Under each subheading, recurring themes that emerge across participants, which influence their well-being, are discussed. Some supporting mechanisms for RAS's well-being under personal and social resources are also stated.

The backgrounds[1] are in the Appendix A. It provides snapshots of some interviewees' and other stakeholders' lives and the extent of their well-being in Hong Kong. This helps contextualize the data presented in the chapters and the findings. Appendices B until F include the data collection tools such as the survey questions, interview, and observation guidelines.

4.1.1 Material Resources

These resources are related to assets with practical use or value, such as land, property, savings, and employment.

- **Housing**

The survey results showed that from RAS perspective housing was the most critical factor influencing their well-being because it determines many other aspects of their lives.

1. Four top issues influencing wellbeing of refugees are: 1. Housing, 2. Rising prices/cost of living, 3. Unemployment, 4. Assistance from the ISS Bureau HK (Table 4.1).

The finding suggests that the high cost of living, the (little financial) help they receive from the ISS, housing, and unemployment are the main factors that influence RAS lives. In addition to that, the interview data shows that the impact of housing extends beyond securing a roof over one's head. The effect of housing is reflected in other areas of life, such as access to good education services. Also, the small size of apartments, the rent and maintenance-related expenses, rising costs may indicate the need for employment.

> Housing determines a lot in Hong Kong. If you go to hospital they ask you for address before they give you services. If you have to take a child to school, they have to see your address…My son had to miss school, he's supposed to be in K-3. But he missed school one year because I didn't have an address. Ms. Iris (Interview)
>
> It's very small. Here is my bed, this is the kitchen, and that is the toilet…it's so stressful, it's like you don't have peace of mind. I mean when you see your daughter wants to play and don't have enough space… Magnolia (the daughter) doesn't know what's my situation,

[1] The data was collected in late 2021; therefore, the descriptions of the participants are provided based on the time of the interview.

Table 4.1 Survey result Q.1 (post-workshop)

#		Count
1	Housing	7.00
2	Unemployment	6.00
3	Rising prices/cost of living	7.00
4	Environmental problems (e.g. air quality, wastes, etc.)	4.00
5	Health provision	3.00
6	Education system	1.00
7	Economic situation	1.00
8	Assistance by the ISS Bureau HK	6.00
9	Duration of claim check/status approval	2.00
10	Taxation	0.00
11	Crime	0.00
12	Political instability (e.g. protests)	0.00
13	None	0.00
14	Don't know	0.00
15	Other (please specify)	0.00

Q1. What are the four most important issues you are personally facing at the moment?

> so she's like 'Why we are living like this? Why we cannot get a big one? A lot of questions (as she is) getting older. Ms. Rose (Interview)
>
> The housing prices they are giving refugees, at the same housing, the prices they are giving to not even refugees. They make it more, double. Even a resident here is looking for rent. They will give you a 3000 house which is normal, but if you are a refugee, they'll make it 6000. Ms. Iris (Interview)
>
> We have our own house[2] now, but I need to pay extra also every month because me and my daughter have only 3900. And I am paying in my house is, uh, 4200 (HKD), so I still have 300 balance and then not included also our electricity and water. So like 600 to 700, I need to prepare every month. Ms. Lilly (Interview)

The third comment may also be considered a reason and/or outcome of social resources as the doubling of rent indicates an intricate interaction between the refugee community and the society.

- **Unemployment**

Refugees are not allowed to work in Hong Kong. Living in this expensive city without a source of income influences their day-to-day well-being. Therefore, refugees and asylum seekers tend to accept any work—full-time, part-time, or dirty jobs. In addition, not allowing legal employment is perceived to encourage refugees to seek illegal pathways.

> The kind of help they give us is as if they're sending us to crime indirectly, because if they give you 1800 for your housing and houses here in Hong Kong, the smallest way I, one person, can enter, not turn his back the other side is 3000 (HKD). What does it mean? What

[2] Lilly stayed in the shelter before due to abuse from her partner, and these shelters are temporary. Hence, 'own house'.

does it mean you who is not working? It means go out and find a way to survive, right? And how do you survive when you cannot work?... People are just trying to survive, and find themselves into crime. Ms. Iris (Interview)

Being a refugee is not so easy and there's some treatment that is not fair for us. And the cost of living is not, the way they give us the support is not really enough for me and my daughter…and they don't allow us to work. They (the government) should give us a legal work (permit), at least even a few hours a day just to have us our own money, not depending on their giving. Ms. Rose (Interview)

Although they expressed the need to find ways to make a living, they also fear getting caught because if they get caught, they go to jail. If they go to jail, who takes care of the child?

Last time I go inside the jail and then my friend take care of my son, because we're not allowed to (bring) the children inside the jail if five years or four years (old). Only one year because you're breastfeeding, you can stay together with your child, but 3, 4, 5 (year olds) cannot go in. Ms. Daisy (Interview)

Although refugees mentioned the need to work and the desire to be independent of government donations, the NGO representative and the SE founder, who is a refugee, reported that the situation concerning employment has improved over the years.

They still don't have the right to work, but now they got the chance. If your case is accepted, you get the chance to apply for work permit. So that's a change. We didn't have that two decades ago... It's complicated. It really scares people off, especially when they've (refugees) crossed the red line without even knowing, and that's why with employment for instance, now we're giving out pamphlets, we're telling employers 'OK, here is step 1–2–3-4-5. That's how we properly and legally get a refugee or a substantiated claimant. Mr. Aaron (Interview)

Right now the government is very nice to refugees. I know people don't believe this but when we came (seven years ago), it wasn't really that. Mr. Indigo (Interview)

We also questioned the NGO representative about the refugees' skill sets to understand whether this could be helpful to their employment. Mr. Aaron explained that the group has a broad composition, from academics to illiterate people. As a result, a wide range of skills and abilities are found in this population. However, he added that refugees' skill sets are not typical; there are many local people with similar skill sets. In an environment where many people are laid off, and businesses are closing down,[3] employers in Hong Kong society may feel uneasy about hiring refugees instead of locals.

When questioned about the reason for being very strict with employment, Mr. Aaron explained how the immigration department deals with employment issues and clarified the department's concerns. From the government's perspective, their strategies and policies are designed to protect employees (in this case, refugees) from exploitation.

For instance, where do you work? Is it one venue or is it multiple venues? If it's multiple venues, it's not going to be approved. Of course they will tell you to change it before they disapprove you, so they want to approve it, but they will really want to make sure that all their concerns are lifted. So, what sort of company it is like? Does it have a good record? So, they're looking for different background information that the company can provide as

[3] The period during and following COVID-19.

> well as a company with good records. Is it financially OK to employ this new person? So, it's a bit of like immigration concerns, but at the same time at least in terms of the criteria, trying to safeguard the employed persons' rights and benefits. So, all that have to be cleared…Once you can clear it off and you start to understand where the immigration is coming from…I personally have not come across a rejected work permit. Of course, everything has to be really neatly dealt with the help of a lawyer. Usually like to gather all the documents and answer questions of the immigration, there is a process. Mr. Aaron (Interview)

4.1.2 Social Resources

Social resources are networks, social ties, and any available structures, such as the formal and informal institutions that regulate one's relationship with society.

- **Lack of recognition**

Findings indicated that a lack of status and lengthy waiting periods influenced refugees' well-being in Hong Kong. The court process is complex and lengthy and can take years or decades. The uncertainty of their future and migration status means the participants in this study experience considerable stress.

> I'm stressed from the judiciary, the High Court, because we don't know yet the decision of the High Court if to go back to Philippines… I am worried if I go back to Philippines…because you know the government is not good there…we have a good hospital, we have a good school for Florent (her son) here. In the Philippines, no, (the government is) corrupt and also people, they're poor." Ms. Daisy (Interview)
>
> "Specially, during court, our case, that one about asylum seekers, it is a little bit stressful. That one if we are going to finish, we need to apply from them because we need cash actually to pay in the high court (for appealing). Ms. Lilly (Interview)

The NGO representative acknowledges the difficulties and barriers this group faces in terms of living a dignified life. He mentioned that the complexity and duration of legal issues are a daily challenge for refugees.

> All of them are going through a screening process and legal support is always needed, and some of them are just totally confused. Even as staff, we spend so much time and effort in asking for professionals and staff to get the hang of, just the flow of it. It's that complicated… We identified them (refugees' needs) as these are systemic challenges that they're facing or the daily challenges they're facing in so many different aspects and that can linger on for years or decades. Mr. Aaron (Interview)

- **Problems with the ISS Hong Kong Bureau**

Problems with the International Social Service (ISS) Hong Kong Bureau, the contracted service provider to claimants in HK, are also heavily related to material resources; however, because the government contracts the ISS as a mediator between the government and refugees, the related problems are considered here under social resources. According to a participant's account, there appears to be miscommunication and/or mistrust between the ISS and refugees.

In addition, the in-kind assistance is reported to be inadequate for maintaining their basic needs and lives. The ISS provides items such as shampoos and detergents, which are also described as low quality. As a result, they prefer not to use them, so they are wasted.

> The way the things are going on between, like personally, me and the ISS is really making life more difficult because you'll have a caseworker who be working not to understand you…You want to tell him how you feel or your well being, and he will be deciding not to understand you. And then there's no way you can help yourself because he's supposed to, you are supposed to be in good communication and understanding each other, to help each other. For him to help me be well and for him, also me, as I help him for his work, right?…we don't know what they (case workers at ISS) report. Maybe they, because they are the ones who are talking to the people...They can put any report on you and because me, I'm so minor that no one would ever know. I'm hidden there. So whatever they want to put on me, they can fool. But I promise you there are so many people who even sleeping on the streets just because ISS is making life miserable for them. Ms. Iris (Interview)
>
> We also have our own personal needs. They only give us the money card, food card that we call, and that's only for frozen bread (food) at ParknShop, not really selling fresh food. So we don't get enough nice food, mostly frozen…They give us one sanitary pads and one (bottle of) shampoo that is not really good shampoo. Per month, one bottle of shampoo and then two pieces of soap and laundry detergent. Ms. Rose (Interview)
>
> In our daily needs in our home like kitchen utensils, like shampoo, Clorox, bleach, something like that and then for cleaning materials, they are also not provided…the shampoo is really the cheapest quality that maybe if you use it then your hair will become very, very hard. So usually I'm not using it and they also give the laundry (detergent) but it's only one pack and it's also like a powder. It's not really cleaning the stains, it's just bubble…I mean the quality is not really good. Ms. Peony (Interview)

- **Limited interaction with Hong Kong society**

The survey results indicated that participants feel they are a part of Hong Kong society (Fig. 4.1).

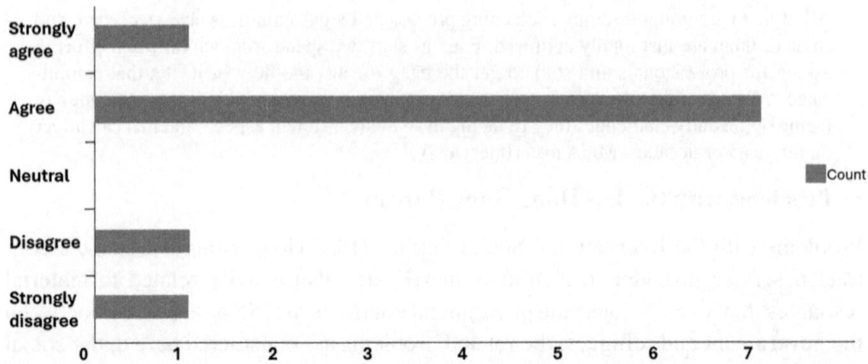

Fig. 4.1 Survey result Q.3 (pre-workshop). Q3—I am a part of the Hong Kong society

However, Mr. Aaron's comments suggest that, despite feeling part of it, refugees often feel isolated from society. He mentioned that media discourse about RAS is negative.

> They're isolated from society and we don't see them on the news unless something bad happens…The portrayal on media and all that, they are just really bad words…even by officials saying that they're just dependent on the system to say the least, using up the resources we have here, while in fact they are barred from doing it. They're barred from generating their own income, unleashing their potentials. They're not allowed to do that, but they commented on to be relying on the system. Mr. Aaron (Interview)

Society's perceptions about refugees were also reported to involve a lot of misconceptions.

Ms. Rose's experience confirms the general lack of knowledge about the refugee community in Hong Kong: "*What kind of people is this or what did they do? Why (are) they like that? So there are big question in their minds… In Hong Kong, maybe some people (know about us), maybe some people don't care.*" (Ms. Rose—Interview).

4.1.3 Personal Resources

Personal resources relate to one's psychology and physiology, including family/friend support, health status, religion, resilience, and hope.

- **Abusive partner**

The interviewees for our study repeatedly shared experiences with abusive partners. The abusive partner problem influences their housing situation, as women reported having to run away from home and stay in a shelter. The Welfare Department in Hong Kong provides these shelters for temporary stays only. After the women leave the shelter, they must find an apartment, but some landlords ask for a deposit. It was reported that the ISS would only pay one such deposit. Therefore, the women reported to face difficulties finding an apartment.

> I have actually problem about her (the child's) father, because we are not married, and he is here. He is also the same, asylum seeker. Then also, (because of him) how many times also living in the shelter actually, because he always do wrong with me, he always beat me.. I told him, I said "you are the only one who beat me like that"…the deposit of the ISS, that one, they give it to us. It's gone already because how many times already I go in shelter because of what the father doing to me. So I need to find a room after the shelter because we cannot stay in the shelter (more than 3 months)…the place is very far going to school, two hours travel. So we go out there very early in the morning…And then I don't have place to stay. So I usually stay in refugee union office (a local NGO), like that. I'm always there. And then I wait until my daughter finish school. Ms. Lilly (Interview)
>
> Her father is, you know, like using drugs. Then he abused me. Then I needed to run away…So I want to change room but I cannot change room…some landlords, they are asking for the deposit...the ISS will only pay one time (deposit). Ms. Peony (Interview)

- **Experience with health services**

The findings suggest that refugees have access to health services and may visit hospitals from time to time. For example, Ms. Iris mentioned she experienced shortness of breath and developed a skin rash during her pregnancy, and Ms. Daisy's son is undertaking speech therapy. However, refugees reported being stressed and abused, but none of the interviewees mentioned receiving or looking for support for abuse or mental health.

> Thanks to God, we are healthy so we are not really going to hospital because in hospital we have also social welfare that they will pay our fees. But if you have a special problem like the eye, then you need pay for yourself because social welfare will not pay for the specialists. Ms. Peony (Interview)
>
> Every time, we go to the hospital on Monday. We go to check out for him (the son) because he has asthma, and then he has speech therapy. Ms. Daisy (Interview)

Rose's experience also indicates that officers in state institutions, such as hospitals or schools, may not understand refugees' documentation/situation.

> Sometimes in the hospital, there were some nurses… if they see we're holding that card,[4] (they ask) What is this? They don't really know what (it is). They say (it is their) first time to encounter. Maybe (they are) new nurses, but sometimes just not really very helpful. Ms. Rose (Interview)

Although some participants reported being physically abused and mentally stressed, they found comfort in having a child, as well as a religion and faith to provide them some peace and support. Since these are individual choices and impact the physiology and psychology of participants, they are categorized under personal resources.

- **Having a child**

Participants indicated they adapt their schedules according to their children's schooling or activities. Therefore, having a child was essential for regulating their lives and providing them with hope.

In addition, allowing children to attend our workshops was viewed positively because refugees have nowhere to leave their children.

- **Religion and faith**

> I really don't know which areas would be improved (about living in HK), but I just I just pray to God that because he's the only person who can satisfy. No worldly things can satisfy a human being. Because you can find a rich people, they will tell you they are not satisfied about some stuff, right? Even with a lot of money, even though some people were in their

[4] RAS have a recognizance document. A Recognizance document is not a visa and simply provides claimants with conditional release from detention pending the resolution of their case. The Recognizance document does not legalize their stay in Hong Kong or confer any rights on them. It also requires them to report to the immigration authorities on a periodic basis (UNHCR Hong Kong, 2025).

> home country, they are still not satisfied. So this thing (being satisfied) I think is more spiritual than what people would think. That what do you look for satisfaction? In this world, if they ask you, you don't have an answer for that. Yeah, so I think you find your satisfaction in God. Ms. Iris (Interview)
>
> The problem also is like if we exceed our utility bills, then we need to pay by ourselves. So that's why when we don't have money we can, we will go to the church and ask for their assistance. Ms. Peony (Interview)

4.1.4 Cultural Resources

Culture relates to gaining knowledge, skills, and values and applying these to life. Cultural resources are shared among members of different small groups within large groups and involve education, training, and community support.

- **Access to education and training**

When participants were asked about access to education, their primary concerns were related to their children's schooling, not their own access to education and training. They said they would like to remain in Hong Kong so their children can complete their schooling because the quality of education back home is poor. In addition, the children may not be able to speak the local language if they return home.

Ms. Iris's son, Jared, had to miss school as they had no fixed address; however, findings indicate that the other participants were satisfied with their children's education in Hong Kong. Ms. Lilly, for example, stated she wanted to stay in Hong Kong for her children's schooling because there is a lack of quality schools in the Philippines, and her children don't know the language.

> I feel better in here because my kids can also go to school right here... If I go back home, the Philippines, they (children) can go school, but it's not the same... Now we have here; it is more easy because we have at least (a list?) of things. The Philippines still don't have because of the pandemic work. And even though the pandemic is over, for example, they need to go to school. Then how? Because they are born here, it may be difficult for them to adjust, especially to the language. Here, they speak in English, they don't know our language. Ms. Lilly (Interview)

Ms. Daisy mentioned being unable to help with Florentin's schooling and feeling stressed by this.

> Because before when he don't have a tutor, Florentin was lazy to do homework because he's a P-3 now. He has a lot of homework and in much Chinese, I don't understand...Every time I only shouting at him because he's not listening to the homework... I don't want to be stressed every time with Florentin because sometimes, my head is eating (me) every time I'm angry to him. Ms. Daisy (Interview)

In relation to access formal education and training, a couple of interviewees explained they had left university due to financial reasons. So, we asked whether they would like to continue higher education. They identified financial issues and time and time constraints as barriers to further study.

> So they offer me that (online university degree)… but the problem was I needed to finish the online course first but then I didn't have much time so I didn't finish it." Ms. Peony (Interview)
>
> I'm not sure because you need to pay by yourself. You need to pay even the kids. If they go to university, you need to pay by yourself already…Under only high school you can give. Ms. Lilly (Interview)

Mr. Aaron also highlighted the difficulties faced by refugees in Hong Kong who want to secure a place in higher education.

> The right to education, to just go to school, it's good that for primaries, kindergarten, primary school and secondary school. They're covered, but beyond that, there's no particular support that can get them anywhere. So if you graduate with straight A's, unless you get scholarship from somewhere or sponsorship from somewhere, that's the end of it…you're stuck in Hong Kong…we've got straight A students…they've got loads of potentials, but once they see that there's no future in that pathway, they start to lose hope. They just (think) 'why am I working hard? Anyway, this is not bringing me anywhere.' Mr. Aaron (Interview)

In terms of access to non-formal education and training, some refugees indicated they had participated in workshops organized by different NGOs, such as informative seminars concerning their rights in Hong Kong (organized by Branches of Hope (BOH)—a local organization working with refugees and asylum seekers-) or handcraft workshops (such as hairband making) organized by another local NGO for mothers. Ms. Iris stated that our workshops represented the first opportunity she had had to participate in some form of education for 7 years. She was the participant who requested a certificate of participation so she could show she had done something in Hong Kong after she returned home (Reflection notes on the interview with Iris).

Regarding activities related to greening skills, Ms. Lilly and Ms. Poppy shared that they had been to several beach and park clean-up activities because their children's school had invited parents. So, they were not the targeted group. Participants said that education is essential for learning about the environment and can improve their participation in greening efforts, such as waste management and sustainable consumption. Regarding non-formal education and skills training, such as our workshops, the NGO representative stressed that the correct use of terminology is a sensitive issue from the government's perspective.

> Training does not lead to employment, right? Because these people, many of them, their cases are not accepted, so they were never get legal employment. So what are you training them for? So, that's the line of argument that we have been facing. Not with us our organization, but with partner organization in the field. Mr. Aaron (Interview)

The findings show that NGOs, such as BOH, are willing to collaborate and/or offer activities for refugees and asylum seekers to improve their situation, which is why they collaborated with us for the study.

Despite facing multiple challenges, NGOs and community support networks help refugees survive and positively influence their lives in Hong Kong. NGOs and community support are categorized under cultural resources because within this area, the RAS group can find comfort, and the NGO community contributes to RAS's well-being thanks to its humanitarian approach. Mr. Aaron reported that RAS support for one another within the refugee community developed organically because they are all in the same situation. Also, findings suggest that the donations they receive from society, like clothing and shoes via NGOs, result in the reuse of items and waste reduction.

- **NGO and community support**

> They find some sort of community not necessarily related to their ethnicities and all that, like that they really get along with one another…You know it's just nowhere near enough for them to just get through the day. But to add on that psychological pressure and if you ask me…it's really between them to support one another. And I think that's the beauty of it. And that's how potentially they can get through the days. Really not about the money, it's not. The money part is important, but that support is irreplaceable that they have from heir fellow community members and us in process. Mr. Aaron (Interview)
>
> I feel shy to ask (my friends) actually because they help you already so much like that… maybe they think, I take advantage. So, I don't (ask them), as long as I can find my way first." Ms. Lilly (Interview)
>
> …in the donation, they can provide some clothes, but people have different sizes, so sometimes I cannot find my size, so usually I cannot find what I really need, the clothes, the shoes. Ms. Peony (Interview)

Thus, based on the survey and interview results, we can conclude that refugees heavily reflected on the lack of available materials and social resources, and how this lack influences their use of or access to other resources. For instance, a lack of housing influences access to education. Nonetheless, their survival mechanisms, such as community assistance and having children, help them deal with daily stress (Table 4.2).

In summary, we found that the participants' perception of well-being stems from a comparative perspective, as their conditions in Hong Kong are perceived as better than those in their home country. They are generally 'fine', which can be expected since they sought asylum in Hong Kong due to the poor conditions in their home countries. Nonetheless, they still shared some problems that influenced their well-being in Hong Kong. Our findings indicate that their agentic capacity is limited due to a lack of resources, including access to employment, healthcare, housing, and education, as well as unhealthy interactions with Hong Kong society and the International Social Service (ISS), Hong Kong Branch. These findings are consistent with issues identified in other refugee-related research in Hong Kong, as explored in Chap. 2 (e.g., Kennedy, Wong, & Ng, 2019). Since their needs directly relate to the government's approach towards them at the societal level, we examine Hong Kong's response to refugees and asylum seekers, consider the initial humanitarian considerations, and discuss the reasons behind the current approach below.

Table 4.2 Summary of findings about resources

Resources		Supportive mechanisms
Material resources	**Housing** *'The housing prices they are giving refugees, at the same housing, the prices they are giving to not even refugees. They make it more, double' Iris* **Unemployment** *'They (the government) should give us a legal work (permit).' Rose*	
Social resources	**Lack of recognition** *'I'm stressed from the judiciary.' Daisy* **Problems with the ISS Hong Kong bureau** *'You'll have a caseworker who be working not to understand you' iris* **Lack of interaction with Hong Kong society** *'In Hong Kong, maybe some people (know about us), maybe some people don't care.'Rose*	
Personal resources	**Abusive partner** *'(because of him) how many times also living in the shelter 'Lilly* **Experience with health services** *'In hospital we have also social welfare that they will pay our fees. But if you have a special problem like the eye, then you need pay for yourself because social welfare will not pay for the specialists." Peony*	**Having a child,** **Religion and faith** *'So this thing (being satisfied) I think is more spiritual than what people would think.' Iris*
Cultural resources	**Access to education and training** *'If they go to university, you need to pay by yourself already…under only high school you can give.' Lilly*	**NGO and community support** *'It's really between them to support one another. And I think that's the beauty of it. And that's how potentially they can get through the days' Mr. Aaron*

4.2 Section 2: Discussion

In the 1970s, Hong Kong welcomed people who fled Mainland China without detaining them and provided them IDs, indicating their welcoming attitude. However, the arrival of Vietnamese refugees in the 1980s and 1990s was difficult (Mathews, 2018). They were detained, and they are still referred to as 'boat people' locally, which reveals the negative image people in Hong Kong have. In the wave after 9/11, we see increased asylee applications from South Asia and some African countries. Their representation in the media varies from 'fake refugees' to 'criminals' who try to abuse the system (Ng et al., 2018). Furthermore, political figures' speeches and media coverage negatively influence society's attitude and create hostility (Justice Centre HK, 2016).

The change in Hong Kong's attitude toward RAS appears to be paralleled worldwide. Crisp (2003) analyzes the shift in the refugee paradigm in developing and industrialized states. According to the analysis, in the 1960s and 1970s, the number of refugees was modest, and they were welcomed, especially in developing countries, because international aid from industrialized and previously colonized countries was made available. The aid was generally channeled through UNHCR to educate, feed, and provide healthcare to exiled people (Crisp, 2003). This aid assisted Western countries' efforts to contain the spread of communism. However, during the 1980s and 1990s, countries such as Malawi, Pakistan, and Guinea were reported to claim their previous commitments to host refugees were easily forgotten by the international community. From this point, media attention and international aid declined. Therefore, countries were left alone to deal with the social, economic, and environmental impacts of influxes of refugees (Crisp, 2003). Following the end of the Cold War and 9/11, refugees were viewed with fear; they were considered to be potential terrorists and received resentment from local people (Mathews, 2018).

The change in welcoming attitudes is attributed to multiple factors:

(i) Increased nationalistic and xenophobic political movements in developing states, which blame refugees for the insecurity and economic hardship local people feel;
(ii) Globalization, terrorism, and racism. Although globalization promotes the mobility of goods, services, products, and highly skilled people, the process disregards the movement of the RAS population and considers them low-level. Underneath this sentiment lies the uncomfortable reality of racism in developed states;
(iii) Lack of public policy, media, and leadership promotes negative attitudes toward RAS groups and results in more authoritarian measures, such as stricter immigration control, rather than improved human rights (Crisp, 2003).

Balzacq et al. (2015) consider these actions to be tools for securitization, which suggests that security issues result from leaders' efforts to understand and shape the world. Such practices use emotions, analogies, stereotypes, etc., in a context where actors (such as policymakers) manipulate the audience's (such as the public or voters) feelings, sensations, and thoughts about an object (which is threatening) in a way that is deemed appropriate by the very actors who created the security issue in the first place (Balzacq et al., 2015). When looking specifically at migration and securitization, Bigo (2001) argues that the media and authorities create a rhetoric of insecurities or dangers (either within their borders or external to them) that results from migration to come up with a 'solution' in the public eye. They often link forms of violence, risk, or threat (such as changes in social norms, behaviors, or the creation of new health risks) with the migrant. For example, "migrants coming from the East" are a danger to Western societies because of their past, which encourages disrespect of the law and criminalization. They are blamed for car theft and any trafficking, including nuclear devices, arms, drugs, or women and children. Migrants from the South are even worse. They are religious fanatics and terrorists, and they try to use welfare to be there without work (Bigo, 2001, p. 4). However, when migration (or other transnational issues, such as drug and human trafficking) is considered and handled by state actors through the lens of securitization, other possibilities and remedies are ignored, such as inequalities or globalization (Balzacq et al., 2015).

For example, by not providing access to TVET programs or employment for RAS's development, the HKSARG contributes to: (i) reducing the possibility of creating more links between refugees and its formal institutions while simultaneously creating a feeling of dependency and loss of agentic capacity for refugees, and (ii) adding to the sentiment of competition and securitization of jobs between locals and RAS. In either case, the distance between RAS and society grows, and RAS face unequal treatment.

Hong Kong's provision of access to such education for RAS is restricted. This situation relates to social justice and equality. However, greening is supposed to alleviate people from poverty and ensure adequate access to resources and quality education. Despite significant attempts at greening in Hong Kong, current practices in Hong Kong do not indicate a full commitment because they do not allow all groups of society, including refugees and asylum seekers, to benefit from its impact. "*A genuine approach to skills for sustainability must be concerned with poverty, inequality, and injustice and contribute to their eradication. It must support individuals' agency whilst also reflecting a careful reading of the structures that too often constrain them. In doing all this, it must minimize the costs and risks of any transformation for the poor and seek to facilitate them into better individual and communal lives, not limit their opportunities for these*" (McGrath & Powell, 2016, p.18).

This brings us to consider how equality is conceptualized in Hong Kong.

4.2.1 Conceptualization of Equality in Hong Kong

Kennedy (2011) explains that Hong Kong is unique in conceptualizing justice and fairness because it has a Western colonial history and a commitment to traditional Chinese values based on Confucianism. The 'rule of law' and an independent judiciary that Hong Kong inherited are considered the influence of its colonial past (Kennedy, 2011). However, by including Confucianist ideas, the suggestion is that.

> Political rule should be impartial or fair (gong in Chinese) to everyone - by that, it means political rule should promote the good of everyone without prejudice or favoritism. In other words, it would be a violation of fairness or justice (gong) if the ruler were selectively concerned about some people only (Kennedy, 2011, p. 166).

The Confucian view of 'fairness' relates to social harmony and is considered more important than individual needs or treating everyone equally (Yung, 2007). However, in the West, social justice relates to treating all equally. In addition, according to Confucianism, a state's intervention in providing equal opportunities is considered inappropriate or unjust because this is a family's responsibility, and they can provide their children with unique advantages such as property (Yung, 2007). However, from a Western perspective, governments can sometimes intervene to ensure everyone can access the same opportunities (Kennedy, 2011). So, according to our understanding, while the Western idea of justice relates more to equality of opportunities, Chinese Confucianism leans towards equality of results. However, for the RAS community in Hong Kong, neither results nor opportunities lead to equality of any kind.

Law and Lee (2013) argue that with the economic rise of Hong Kong and a return to the Mainland, a new identity, 'Chinese Orientalism', has emerged in the city. This identity is a merger of British colonialism and traditional Chinese racism and is insensitive towards minority groups (Law & Lee, 2013). It has also been suggested that there is no acknowledgment of the value of diversity in Hong Kong. Therefore, it seems that policies related to justice and fairness, e.g., antidiscrimination ordinances, are pursued as a pragmatic policy objective. As a result, only concessions that are deemed reasonable by the bureaucracy and the government are granted (Kennedy, 2011, p. 166). Studies also indicate that bureaucratic and government leaders can justify these policies or their approach by using a discourse about securitization (Balzacq et al., 2015; Bigo, 2001; Wood & Flinders, 2014). Although there have been improvements concerning policies and practices regarding refugees over the years, Mathews (2018) argues that there is a growing awareness and more welcoming attitude towards the RAS compared with before. However, the reason behind this is not a concern for asylum per se but is the result of anti-Chinese sentiment in the city.

Our interaction with RAS and NGO representatives confirmed that HKSARG has a pragmatic attitude towards this group only in terms of adherence to international legislation, such as the International Covenant on Civil and Political Rights (ICCPR) and the provision of international protection, as the number of refugees is not large enough to have an impact racially or ethnically in the city (Race Relations Unit, 2022). Therefore, the government can disregard refugees without experiencing severe community backlash. However, the government must still be held accountable for its actions and convince the community (Kennedy, 2011) because of the 'rule of law' that distinguishes it from Mainland China. Studies suggest that if the UN Refugee Convention were to be signed, there would be a legal basis for refugees to hold the government responsible for providing the conditions it does not provide at the moment, such as recognition and employment (Loper, 2010; Kennedy, Wong, & Ng, 2016). The government's strict stance on not signing the Convention and previous remarks from the city's executive leadership to withdraw from the UN Convention of Torture 'if needed' (Ngo & Leung, 2016) suggests that Hong Kong does not want to recognize RAS and refrains from endorsing policies that bind the government to the provision of certain rights for this group.

For these reasons, we believe Hong Kong's strategy de-politicizes the refugees.

In essence, politicization means making a matter the subject of public policymaking and/or public discussion and political conflict, whereas depoliticization implies the opposite (Feindt et al., 2021). Flinders and Buller define depoliticization as "the range of tools, mechanisms and institutions through which politicians can attempt to move to an indirect governing relationship and/or seek to persuade the demos that they can no longer reasonably be held responsible for a certain issue, policy field or specific decision" (Flinders & Buller, 2006). This type of strategy is regarded as 'narrow' in its focus because it is very state-centric (Jenkins, 2011) and does not entail multi-level analyses or actions of other actors in policymaking (Wood, 2016). However, we adopt this narrow definition for the following reasons:

1. As articulated by Lai and Kennedy (2017), some social actors, such as NGOs and lawyers, have a positive attitude toward this RAS, so it is the policymakers and media whose attitudes are criticized internationally. Therefore, the essence is narrow.
2. This is not a political study but rather an educational one. However, we found issues influencing refugees' well-being largely related to a lack of policies and practices that address their needs. On this note, the current literature regarding policy review for RAS is limited and lacks an appropriate term to describe/assess the policies (Choy & Shi, 2021).

As we aim to develop recommendations about improving the well-being of refugees through their empowerment and addressing external constraints, we discuss how the RAS agenda can be politicized.

4.2.2 *Depoliticization of RAS*

Wood (2016) reports that historically, following the win of liberal democracy over Soviet communism in the 1990s, a wave of privatization and globalization was seen in liberal democratic states, which was accompanied by declining levels of interest and engagement in politics (Wood, 2016). During this period, what became important was utility maximization, self-interest, and the superiority of the market in all social areas, also known as neo-liberalism (Wood, 2016). Being a free-market economy, the impact of this wave can also be observed in some policies in Hong Kong, such as in education. One example is the increased number of international schools in the city because the government allocated vacant school premises or greenfield sites to international school operators to build new schools with the hope of further globalization (Woo, 2014).

Flinders and Buller (2006) suggest that the rationale behind depoliticization is to reduce politicians' direct control by delegating or minimizing discretion, which in turn helps to alter market expectations regarding the credibility of policymaking from a public policy perspective. Also, adopting an indirect governing relationship reduces political overload, and in case of a policy failure, they (policymakers) can shift the blame onto other groups. Wood and Flinders (2014) introduce three faces of depoliticization:

- Governmental depoliticization: focuses on the transfer of issues from the governmental sphere to the public sphere through the 'delegation' of those issues by

politicians to arm's-length bodies, judicial structures, or technocratic rule-based systems that limit the discretion
- Societal depoliticization: involves the transition of issues from the public sphere to the private sphere and focuses on the existence of choice, capacity deliberation, and the shift towards individualized responses to collective social challenges
- Discursive depoliticization: involves transferring issues from the private realm to the 'realm of necessity' where 'things just happen,' and contingency is absent. It, therefore, focuses on the role of language and ideas to depoliticize certain issues and, through this, defines them as little more as elements of fate (p.165).

The literature review and findings show that discursive depoliticization is apparent in Hong Kong. The RAS group has limited interaction with society, which lacks knowledge about them, and the media creates negative images of the RAS community. In addition, the community encounters negative attitudes from society and public authorities such as hospitals or ISS. This feeling of being overlooked stems from a lack of coverage and explanation for why and how this group lives in Hong Kong.

The media utilizes language about RAS, such as 'fake' and 'criminals', the locals charge refugees extra for housing, the authorities neglect to understand them, and policies do not allow them to earn a living. There is also the question of 'real' or 'fake' refugees, which falls out of the scope of our study. However, we understand that it is important for the city, and even within the refugee community (Mathews, 2018), to protect the system from abusers. Therefore, attempting to implement increased border controls may be deemed essential. However, according to Bigo (2001), such implementation

> creates rhetoric about the protection of the borders as a fundamental of national sovereignty, which is very far from the actual practices of controls of the borders. Governments want to show that they are tackling the situation at the borders and asking the security agencies for "results". But thereafter, the approach obliges the politicians to accept responsibility of the arrival of illegal immigrants, at a time when they have not the power to block them, even if they want to. It confronts politicians with the reality of the limits of their effective power and their refusal to accept that immigration…The result of this is that the local mayor sees the undocumented migrant as his personal enemy, as an affront to his authority (p. 9).

This narrow and negative focus makes this group political/news material to be used when necessary. For this reason, the refugee group feels trapped by the system and within it, and society conceptualizes the group's problems as external to them. It is not clear whether the government or society knows about their situation. However, the problems are internal and result from a lack of recognition and a denial of their potential social function by the authorities, their interaction with society, and participation in social areas such as jobs or educational institutes. Therefore, discursive depoliticization forms '*necessities, permanence, immobility, closure, and fatalism… concealing, negating or removing contingency*' (Jenkins, 2011, p. 160) within Hong Kong.

There is a fear of attracting more refugees into Hong Kong and turning it into a refugee heaven, which may create resource competition. Turning the focus from humanitarianism into a matter of 'security' and identifying an existential threat for different aspects of the social order (e.g., threats to the job market and housing) is a tool for depoliticizing (Wood & Flinders, 2014). Therefore, we understand that this depoliticizing act increases inequalities and asymmetries in society and negates refugees' efforts to move beyond their dismal lives and contribute to society.

Based on the argument above, political reasons supported humanitarian action for certain groups in Hong Kong's history, such as the Chinese in the 1950s or the Vietnamese in the 1970s. However, the RAS group is currently composed of mixed backgrounds with low acceptance rates. Then, is it better for the government to depoliticize its actions?

The government must differentiate between the depoliticization of humanitarian action and the depoliticization of refugees as a group. Depoliticization of humanitarian action is necessary and supported by the UNHCR (Grandi, 2021) because if governments are 'of the people, by the people, and for the people,' they should protect the dignity of these people because this is the purpose of a government. However, in Hong Kong, humanitarian assistance is provided at a minimum level via a contracted NGO, and refugees are destitute. On the other hand, the RAS community is removed from the public sphere and subject to judgmental language in the media. This experience of negative valuation and discrimination within social institutions because they are a part of a stigmatized group results in the disempowerment of the group (Solomon, 1987). It should be up to people to decide whether they want to be depoliticized. If people want their voices heard and/or seen and represented on a political platform, that should be endorsed. If they opt not to, the choice indicates a level of freedom and adherence to democracy.

Therefore, when addressing this challenge, societies, and governments need to define their relationship with disadvantaged groups because the most important indicators of their development are based on their approaches to civilization, that is, human relations and the value given to human beings and practices related to this. Civilization and development are not simply about protecting the rights and laws of its citizens and securing their basic needs, such as education, health, and shelter (Anderson, 2014). On the contrary, it is to ensure the rights to education, health, shelter, and a humane life by observing the rights and laws of disadvantaged people who, for one reason or another, take refuge in and entrust themselves to the administration, which they consider a safe haven. Those who ignore these rights and claim they are a civilized society, regardless of their level of development, are deceiving themselves. Therefore, humanitarian values should be defended in all aspects of life; they reflect the value attributed to human life and represent a shared cultural heritage created by all societies. Protecting this heritage is an essential outcome of a humane approach. In addition, such an approach should be considered a positive attempt to ensure the integration of disadvantaged groups within society and to stir their conscience and willingness to contribute to sustainable (urban) development.

4.3 Summary

To summarize, our findings on the issue of resources and well-being show that refugee and asylum seekers are willing to contribute to the city yet their access to resources is denied in Hong Kong. The help they receive from the government is not considered enough, especially considering the increasing prices and lack of access to the labor market. Lack of access to certain resources influences other areas. Such as, lack of access to housing influences RAS kids' schooling, and they are stressed. Participants in this study were abused by their partners; however, their children seem to give them hope and resilience for the future. In addition, the support within the RAS community to one another and religious groups is appreciated and helps address some of the needs. However, most of the problems could be eliminated if HKSARG changed its approach to this demographic. After looking at the conceptualization of equality in HK and its approach to refugees and asylum seekers as a community, we found that RAS are depoliticized in Hong Kong. This hampers their agency and limits their capability by the state apparatus. Our analysis of why the state acts in this manner suggests that the formation of its mentality and governing style may be related to Confucian values, its colonial background, and the process of new identity formation in the city. Education and training, whether received in formal or informal settings, are highly valued by this community and greatly benefit them. Thus, actions should be taken to improve their situation, especially through capacity building, so that they can contribute to SUD in Hong Kong.

The next chapter presents findings about the efficacy of the workshops and discusses how they can lead to empowerment, especially for women.

References

Anderson, T. (2014). Human development, the state and participation. *Development Studies Research, 1*(1), 64–74. https://doi.org/10.1080/21665095.2014.933080

Balzacq, T., Léonard, S., & Ruzicka, J. (2015). 'Securitization' revisited: Theory and cases. *International Relations, 30*(4), 494–531. https://doi.org/10.1177/0047117815596590

Bigo, D. (2001). Migration and security. In V. Guiraudon & C. Joppke (Eds.), *Controlling a new migration world.* Routledge.

Choy, W. C., & Shi, X. (2021). A comprehensive review of refugee and asylum studies in contemporary Hong Kong: Law, policies, and lived experiences. *Asian Journal of Social Science, 49*(1), 38–48. https://doi.org/10.1016/j.ajss.2020.11.002

Crisp, J. (2003). *A new asylum paradigm? Globalization, migration and the uncertain future of the international refugee regime. Evaluation and policy analysis unit.* UNHCR.

Feindt, P. H., Schinwindenhammer, S., & Tosun, J. (2021). Politicization, Depoliticization and policy change: A comparative theoretical perspective on Agri-food policy. *Journal of Comparative Policy Analysis: Research and Practice, 23*(5–6), 509–525. Retrieved from https://www.tandfonline.com/doi/full/10.1080/13876988.2020.1785875

Flinders, M., & Buller, J. (2006). Depoliticisation: Principles, tactics and tools. *British Politics, 293-318.* https://doi.org/10.1057/palgrave.bp.4200016

Grandi, F. (2021, December 8). *It's important to depoliticize humanitarian work.* (U. H. Refugees, Interviewer) Retrieved from https://www.youtube.com/watch?v=e3-qBFBAIJI

Jenkins, L. (2011). The difference genealogy makes: Strategies for politicisation or how to extend capacities for autonomy. *Political Studies, 59*(1), 156–174. https://doi.org/10.1111/j.1467-9248.2010.00844.x

Justice Centre HK. (2016, April 11). *Joint statement*. Retrieved from Justice Centre Hong Kong. https://www.justicecentre.org.hk/jointstatement-eng/

Kennedy, K. J. (2011). The "long march" toward multiculturalism in Hong Kong: Supporting ethnic minority students in a Confucian state. In J. Phillion, M. T. Hue, & Y. Wang (Eds.), *Minority students in East Asia: Government policies, school practices, and teacher responses* (pp. 155–173). Routledge.

Kennedy, K., Wong, K. L., & Ng, H. Y. (2016). *Towards a refugee policy for an inclusive Hong Kong: Enhancing the status of China's International City. The education University of Hong Kong, Centre for Governance and Citizenship*. Central Policy Unit of the HKSAR Government. Retrieved from https://www.pico.gov.hk/doc/en/research_report(PDF)/2014_A5_010_15B_Final_Report_Prof_Kennedy.pdf

Kennedy, K. J., Wong, K. L., & Ng, H.-Y. (2019). Being asylum seekers in Hong Kong: Institutions, action and power. *Asian Journal of Social Science, 47*, 534–565. https://doi.org/10.1163/15685314-04704006

Lai, A. P., & Kennedy, K. J. (2017). Refugees and civic stratification: The 'Asian rejection' hypothesis and its implications for protection claimants in Hong Kong. *Asia and Pacific Migration Journal, 26*(2), 206–223. https://doi.org/10.1177/0117196817706173

Law, K.-Y., & Lee, K.-M. (2013). Socio-political Embeddigns of south Asian ethnic minorities' economic situations in Hong Kong. *Journal of Contemporary China, 22*(84), 984–1005. https://doi.org/10.1080/10670564.2013.795312

Loper, K. (2010). Human rights, non-refoulement and the protection of refugees in Hong Kong. *International Journal of Refugee Law, 22*(3), 404–439. Retrieved from https://hub.hku.hk/bitstream/10722/92989/2/Content.pdf

McGrath, S., & Powell, L. (2016). Skills for sustainable development: Transforming vocational education and training beyond 2015. *International Journal of Educational Development, 50*, 12–19.

Mathews, G. (2018). Asylum seekers as symbols of Hong Kong's non-Chineseness. *China Perspectives, 18*(3), 51–58. https://doi.org/10.4000/chinaperspectives.8132

Ng, I., Choi, S. F., & Chan, A. L. (2018). Framing the issue of asylum seekers and refugees for tougher refugee policy–a study of the media's portrayal in post-colonial Hong Kong. *Journal of International Migration and Integration*, 1–25.

Ngo, J., & Leung, C. (2016, January 14). *Hong Kong could quit torture convention, says CY*. Retrieved July 5, 2021, from South China Morning Post. https://www.scmp.com/news/hongkong/politics/article/1901085/hong-kong-could-quit-torture-convention-says-cy

Race Relations Unit. (2022, August 16). *The demographics: Ethnic groups*. Retrieved from Information Centre: https://www.had.gov.hk/rru/english/info/demographics.htm

Ryan, D., Dooley, B., & Benson, C. (2008). Theoretical perspectives on post-migration adaptation and psychological well-being among refugees: Towards a resource-based model. *Journal of Refugee Studies, 21*(1), 1–18.

Sen, A. (1999). *Development as Freedom*. Oxford University Press.

Solomon, B. B. (1987). Empowerment: Social work in oppressed. *Journal of Social Work Practice: Psychotherapeutic Approaches in Health, Welfare and the Community, 2*(4), 79–91. Retrieved from https://www.tandfonline.com/doi/abs/10.1080/02650538708414984

Woo, D. (2014). Neoliberalism in two Hong Kong school categories. *Current Issues in Comparative Education, 16*(1), 37–48.

Wood, M., & Flinders, M. (2014). Rethinking depoliticisation: Beyond the governmental. *Policy & Politics, 42*(2), 151–170. Retrieved from https://www.researchgate.net/publication/263764363_Rethinking_depoliticisation_Beyond_the_governmental/download

Wood, M. (2016). Politicisation, depoliticisation and anti-politics: Towards a multilevel research agenda. *Political Studies Review, 14*(4), 521–533. https://doi.org/10.1111/1478-9302.12074

Yung, B. (2007). An interplay between Western and Confucian concepts of Justice: Development of Hong Kong housing policy. *Housing, Theory and Society, 24*(2), 111–132. https://doi.org/10.1080/16506070601119397

Chapter 5
Empowering Refugee and Asylum-Seeking Communities through Social Entrepreneurship

Abstract This chapter analyses the findings of the study related to the workshops and their effectiveness. They are presented in three sections. The first section reports on the findings related to different aspects of a series of workshops conducted with refugee and asylum-seeking women in Hong Kong such as the need, purpose, principles, and use of ESD teaching and learning (pedagogy) with a particular focus on generic green skills development, so that RAS can support SUD in Hong Kong. The findings in this section are presented according to learning experiences at different stages of the workshops, namely before, during, and after, to evaluate the impact of the intervention on learning (Level 2), behaviour (Level 3), and results (Level 4).

In the second section, we share the findings on empowerment through social entrepreneurship. Empowerment is a change in behaviour and the ability to take control of decisions impacting one's life at the personal or community level. It requires effective use of one's resources. However, this can be influenced by structural issues at the societal level. If and when permitted, establishing a social enterprise could be a vehicle for RAS in Hong Kong to apply the skills they learned, and this, in turn, could have potential impacts at the societal, personal, and community levels because establishing a social enterprise is not only about generating income but also to find solutions for environmental and/or social issues. Therefore, the findings in this section are categorized at the societal, personal, and community levels.

In the third section, we discuss these findings based on an evaluative framework applied for the study (See Chap. 1 for details on the research framework) as well as based on additional literature review concerning studies on refugee women and their empowerment. We conclude that although RAS indicated a lack of knowledge concerning sustainability at the beginning of workshops, their understanding and practices improved and positively changed due to their participation in the workshops. They indicated they valued learning new skills through generic green skills workshops, and they could be applicable at home. Social entrepreneurship is considered an inspiring concept; however, refugees do not have the freedom to establish one or work in one. We extended our discussion to women entrepreneurship and the barriers they face and concluded that participation in social entrepreneurship skills training increased their capability and gave them an option to choose.

A. T. Saral, M. Pavlova, *Education for Refugees and Sustainable Green Urban Development*, Education in the Asia-Pacific Region: Issues, Concerns and Prospects 78, https://doi.org/10.1007/978-981-95-1680-3_5

Keywords Refugees' experience · Skills training · Barriers and benefits before employment for refugees · Programs to enhance refugee women's skills

5.1 Section 1: Findings Related to Different Aspects of the Workshops

We observed and found the following attitudes, knowledge, and skills among our participants based on the results of the survey they completed both at the beginning and the end of workshops, and personal and focus group interviews we held after the workshops. Observation notes were kept during the workshops, and reflections were recorded afterwards.

5.1.1 Before Workshops

- **Environment not a priority, but willing to learn new skills and knowledge**

The participants' first encounter with greening was the Green Refugee Day we organized. The NGO that collaborated with the participants provided positive feedback about this event, and they expressed an interest in the workshops in the following months. Therefore, *Level 1—Reaction* towards our evaluative framework was positive, and the result for the first level was successful.

We asked participants and the NGO representative about their motivation to participate in the workshops and found that environmental issues or sustainability were not topics at the forefront of their minds because they have to deal with survival on a daily basis. However, the refugee community indicated its willingness to acquire new skills and knowledge.

Ms. Iris mentioned that environmental issues were not something they were able to be concerned about in their situation.

> I don't see the environment. How would you say environment when you are not sleeping? …last night you don't have a place to stay, will you say environment? Maybe you can ask that from owners who can understand what is going on. But for us, we are dying. We're in the hospital all the time. I don't even have where to sleep. Then you're asking about the environment. I'll be lying if I tell you anything about the environment. Ms. Iris (Interview)

Despite not prioritizing environmental sustainability, the participants were motivated to learn something new.

Image 5.1—Their motivation was to learn skills and knowledge that are new for them. The level of the reading materials is good, but it may be hard to get the capacity (time, for example) to go through them. The hands-on activities are enjoyed by the participants and easy to engage them. They are glad that the children were involved in activities too. The pace is also good. As for suggestions, they would like more activities to be included in the workshops. Mr. Aaron (Observation notes)

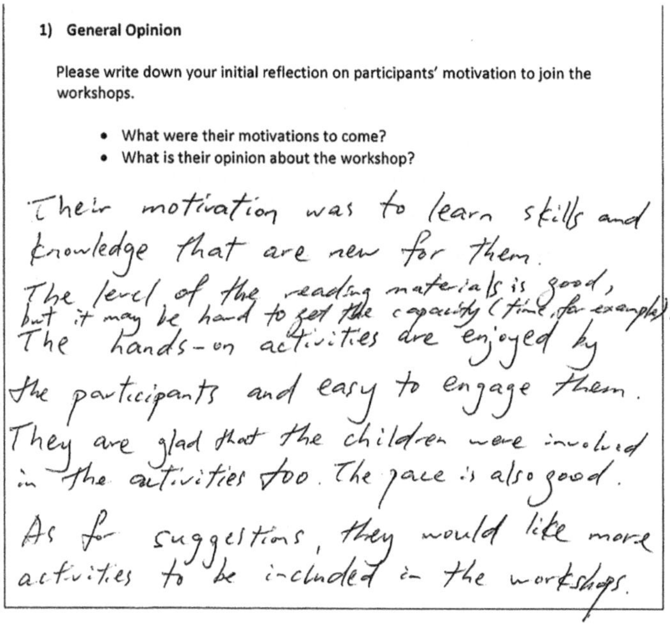

1) **General Opinion**

Please write down your initial reflection on participants' motivation to join the workshops.

- **What were their motivations to come?**
- **What is their opinion about the workshop?**

Their motivation was to learn skills and knowledge that are new for them.
The level of the reading materials is good, but it may be hard to get the capacity (time, for example) to go through them.
The hands-on activities are enjoyed by the participants and easy to engage them.
They are glad that the children were involved in the activities too. The pace is also good.
As for suggestions, they would like more activities to be included in the workshops.

Image 5.1 RAS motivation to participate in workshops—Observation notes by the NGO representative

Similarly, the motivation for the NGO is also to try something new.

> This initiative that you're bringing here with us, like as I said in the beginning, we are not aware of this at all. Beach cleaning (is) maybe one-off like (activity). We do that sometimes, but in terms of lifestyle or something that you can do on a daily basis in your life, it's completely new to many of us. And as an organization we may not be fully aware of that when we are planning our programs. It's not part of our programs because we don't even have that in the back of our head Mr. Aaron (Interview)

- **Low level of awareness and misinformation about waste**

We inquired about the participants' attitudes and knowledge toward waste before the sustainable consumption and waste management workshops.

Participants' reactions indicated a range of zero to low-medium awareness about state and enterprise-level initiatives about waste management. For instance, Ms. Lilly said she had seen the Bauglio's (a green group) tag at different stores while walking, but the others had not noticed it. Another example was the "No-straw" campaign at McDonalds. Most participants said they had seen it but had not paid attention to the labels and did not know what it meant before (i.e., green label).

Ms. Iris commented, "*You know, we don't know these labels, we don't pay attention, so it is new to us.*" The other women in the group nodded in agreement (Reflection memo, Week 2—Sustainable Consumption).

Misinformation was spread about practices they were aware of. For example, they thought the number inside triangles under plastic bottles represented how many times they could be used, and they believed wet tissues could be recycled (Reflection memo, Week 3—Waste management).

We asked participants if they recycle, and the answer was always. When questioned where they picked up this habit, there was an assortment of answers. Some participants said they started doing this in Hong Kong, while others indicated they learned it in their homeland. However, the findings did not support this when we visited one of the participants' houses and talked with others during workshops.

- **Pro-environmental action without conscience**

According to the survey, prior to the workshop, some participants were aware of their energy consumption (item 3), water consumption (item 9), food consumption (item 12), and clothing consumption (item 18) (see Table 5.1).

We asked participants to provide an example from their daily lives related to consumption. Ms. Dahlia commented that she cut up a plastic bottle to make a toothbrush and toothpaste holder in the bathroom, and another participant said she uses a fabric bag. Another participant said she brought her plastic bag to the supermarket. Another gave an example of re-filling a plastic bottle and reusing a tissue paper box after the paper was used. However, they did not consider how these actions would affect or contribute to the environment or the lives of others.

We also asked who they perceived was responsible for solving environmental problems (Table 5.2). They mentioned governments, NGOs, and enterprises as being primarily responsible.

We tried to understand the reasons behind their actions concerning sustainability. The motivation they mentioned behind their mindful consumption was financial rather than environmental.

> No water because I pay 50 (HKD) a month, so I (use) less, less. And then for the electricity, that's the one also, because if too much use I too much pay. So I need to control my electricity, when go out (I turn) off everything. Ms. Lilly (Interview)

Half the participants replied "never" to item 13 in Table 1. For clarification purposes, in this item we asked whether they avoid buying products that pollute the air or whether they never avoid buying those products. The participants replied that they never avoid buying air fresheners or deodorant. Therefore, if they need these items, they buy them without considering the environmental impact.

Some participants left the question about home composting (item 11 in the same table) blank. We asked why and how they could increase their participation in such activities. There was a mix of answers. Ms. Lilly said she was in too much of a hurry; Ms. Iris said she did not know what it was. Ms. Rose, on the other hand, indicated that although she knew how to do it, there was no space or reason for her to compost at home. Therefore, leaving an item blank is due to a lack of time, knowledge, and space (Reflection memo, Week 3—Waste Management). The mix of answers also indicates that participants wanted to express their diverse opinions and felt comfortable sharing them.

From these results, we concluded the following:

Table 5.1 Survey result Q.22 (pre-workshop)

#	Question	Always	3+ times a week	1–3 times a week	1–2 times a month	1–2 times in six months	Never	Total
1	I separate plastics, papers and glasses for recycling.	40.00%	10.00%	20.00%	20.00%	0.00%	10.00%	10
2	I re-use plastics products to avoid single-use (e.g. re-use plastic bags, refill my own water bottle)	50.00%	20.00%	0.00%	30.00%	0.00%	0.00%	10
3	I cut down my energy consumption by turning down air-conditioning, not leaving appliances on stand-by or buying energy-efficient products.	40.00%	20.00%	0.00%	20.00%	0.00%	20.00%	10
4	I speak to others about environmental issues.	0.00%	37.50%	25.00%	12.50%	12.50%	12.50%	8
5	I repair a product instead of replacing it.	22.22%	11.11%	22.22%	22.22%	0.00%	22.22%	9
6	I avoid buying over-packaged goods.	33.33%	22.22%	0.00%	0.00%	11.11%	33.33%	9
7	I buy or use second-hand products instead of new ones such as clothes or electronics.	30.00%	20.00%	0.00%	0.00%	50.00%	0.00%	10
8	I join environmental demonstrations, attend environment-related workshops, take part in environmental activities (e.g. a collective beach or park cleanup)	20.00%	10.00%	10.00%	0.00%	30.00%	30.00%	10
9	I cut down my water consumption such taking showers in less than 5 min or using less water to wash dishes.	60.00%	10.00%	0.00%	20.00%	0.00%	10.00%	10
10	I use rechargeable batteries.	12.50%	12.50%	0.00%	12.50%	12.50%	50.00%	8
11	I make home composting.	0.00%	14.29%	14.29%	14.29%	14.29%	42.86%	7
12	I buy exactly what I need to avoid food waste and other types of wastes.	50.00%	30.00%	0.00%	20.00%	0.00%	0.00%	10

(continued)

Table 5.1 (continued)

#	Question	Always	3+ times a week	1–3 times a week	1–2 times a month	1–2 times in six months	Never	Total
13	I avoid buying products that pollute air such as hair spray and air freshener.	10.00%	20.00%	0.00%	0.00%	20.00%	50.00%	10
14	I warn someone who litters.	42.86%	0.00%	14.29%	0.00%	14.29%	28.57%	7
15	I talk to my kids about protecting the environment.	62.50%	12.50%	0.00%	12.50%	0.00%	12.50%	8
16	I share links on social media concerning environment.	11.11%	0.00%	0.00%	22.22%	22.22%	44.44%	9
17	I follow an environmental organization online (such as UNEP, WWF, etc.).	11.11%	0.00%	11.11%	0.00%	0.00%	77.78%	9
18	I donate/sell the items I no longer need such as clothes, plates, books, etc.	25.00%	12.50%	0.00%	12.50%	25.00%	25.00%	8
19	I dispose of household batteries and light bulbs in designated areas.	30.00%	10.00%	0.00%	20.00%	20.00%	20.00%	10
20	Other (please specify)	0.00%	66.67%	33.33%	0.00%	0.00%	0.00%	3
21	Other (please specify)	33.33%	33.33%	0.00%	33.33%	0.00%	0.00%	3
22	Other (please specify)	0.00%	33.33%	33.33%	33.33%	0.00%	0.00%	3

Q. 22 Please indicate how often you do the activities below (For the "other" option one participant wrote walking (3+ times a week), singing (always), and shaving (1–3 times a week). Although these additional activities may appear irrelevant, they could relate to the sustainability of her well-being from an alternative perspective, however, this was not confirmed.)

Table 5.2 Survey result Q.8 (pre-workshop)

#	Government	Count
2	Private companies/enterprises	9.00
7	Other (please specify)	1.00
3	Non-governmental organizations (i.e. NGOs)	9.00
5	My household	5.00
1	Government	10.00
6	Combination of several (please specify)	6.00
4	All households	8.00

Q8—In your opinion, who is responsible for solving environmental problems? (1—Most responsible, 5—Least responsible) You may rank up to five. You can type your preferred ranking in the text box.

(i) the participants engage in some reuse, reduce, and repurpose activities, but not consciously (Reflection memo—Week 1 Sustainable consumption); and
(ii) the survey results and real-life actions do not always cohere.

5.1.2 During the Workshop

- **Limited understanding of the complexity and connectedness of SD issues**

During the workshop, participants were asked to discuss whether and how environmental problems contribute to their personal problems concerning health, housing, unemployment, and ISS assistance (see Image 5.2).

However, the participants could not immediately identify a connection between the two aspects, namely environmental problems and other areas of concern. When probed individually about how air quality and waste influenced their health or housing, some of them spoke up. For example, Ms. Iris mentioned that due to poor air quality, overconsumption, and inadequate waste management, her apartment was infested with cockroaches and rats. So, these insects and animals may infect her cups and plates because they carry diseases. Because of this, she said she used to visit the hospital every week, especially when she was pregnant with her son. She experienced shortness of breath, stomachaches, and skin rashes. (Reflection memo—Week 3—Waste Management)

- **A sense of community**

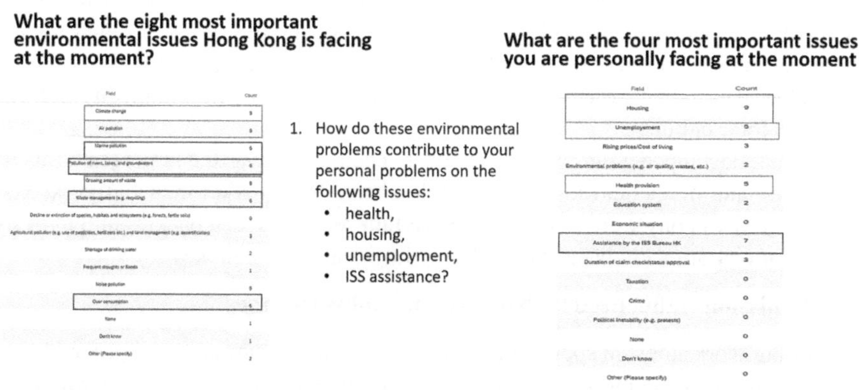

Image 5.2 Discussion points during workshops

The reflection notes indicated that participants formed good relationships and helped one another during the workshops. Ms. Jasmine, a South Asian asylum seeker, asked for help during the toy/wallet making hands-on activity and talked to other participants. She indicated higher communication (interpersonal) skills than others in this workshop. However, this behavior does not necessarily appear in the greening context. So, we noted that it could be channeled into greening via education (Reflection memo—Week 2—Sustainable Consumption).

- **Variety of (business) skills, cultural perspectives, and experimenting**

The participants indicated different levels of (business) skills, which may be related to their previous work experience. For instance, except for one woman, the others reported they had not started an enterprise yet but had worked abroad in different industries.

Most participants did not seem to possess artistic skills except Ms. Jasmine. Ms. Jasmine was exceptionally active in bottle painting and face/body scrub making because she has a background in the cosmetic sector and, at the time of the workshop, had made a deal with a center to make ethnic bracelets, earrings, necklaces, and bangles (Reflection memo—Week 4/5—Social Enterprise I/II).

All participants enjoyed the hands-on activities and took home their produced items. In addition, the women were experimental during hands-on activities. While some were curious about adding colour to the candle mixture, some explored different scents, and some added cinnamon for testing purposes, they were happy to experiment and create independently (Reflection memo—Week 3—Waste Management).

During the Social Enterprise workshops, three groups were formed. Ms. Iris wanted to work alone for the SE-business canvas activity because she said she already had an idea and that the others would not be familiar with the African context. She explained that people in her society have a culture of reuse and repair; however, unless the second-hand clothes are from the US, they are not considered desirable. The other women, participants from the Philippines, said this did not apply to their culture.

Team 2 developed a family store concept. Their group was the largest in terms of members, and they could identify multiple channels to market their business. Ms. Rose and Ms. Peony were in this group, asylum seekers with a background in business administration studies, so they drove this idea (see Table 5.3).

- **SE missions influenced by background and workshops**

During workshops on social enterprises, the business canvas model proved helpful in designing and guiding participants' ideas about business (Reflection Memo—Week 4—Social Enterprise I).

The group members and their respective missions are listed in Table 5.3 (See Appendix F for their canvas' filling).

We recognize that their business ideas were shaped by both their personal experiences and our workshops. For instance, in Ms. Iris's business idea, the mission

Table 5.3 SE teams and missions

Team No.	Group Members	Mission
1	Iris	Establishing a church and a farm. Anyone who comes to the church works on the farm and gets food that day for their work. The food would also be distributed to orphanages, HIV-positive patients, and the elderly.
2	Peony, daisy, rose, dahlia, jasmine	Establishing a family store. The items for sale would be food, clothes, electronic materials, etc. these materials would be collected from others and distributed to those in need.
3	Lilly and Poppy	Making stuffed toys for orphanage children.

involves orphanages because her son was with the Social Welfare Department.[1] For Team 2, items such as food, clothing, and electronic materials would be collected from others and distributed to those in need, including themselves. This relates to their experience of donation and reuse in Hong Kong.

The only group to generate its business idea based on an environmental issue was Team 3. After inquiring about the environmental impact of their mission, the other groups also incorporated the reuse of materials and waste reduction into their objectives. This indicates that they had derived inspiration from our previous workshops and had developed a greater sense of environmental awareness. Distributing to the less fortunate or needy groups can, to a certain extent, indicate social awareness or responsibility.

- **More time to conceptualize SE**

During the second week of the SE workshops, the teams worked on their business canvas before progressing to creating a pitch talk (Please see Appendix F for the team's canvas). Team 3 did not come to the first session of the SE workshop, whereas Team 2 was present both weeks. Team 2 was more detailed in their pitch talk than Team 3. From this perspective, the details in the forms indicate that the thinking and designing require time. Therefore, spending more time on this topic can help identify finance-related areas and fill in the blanks on the business canvas.

5.1.3 After the Workshop

- **Increased awareness and willingness to learn about sustainable development**

Participants' awareness and willingness to learn about sustainability improved. For instance, Ms. Petunia asked us to send her the link for making wallets from fabric so she could try making them at home. She also sent a message on WhatsApp about

[1]Due to her financial situation and abusive partner, she said that the Social Welfare Department would take care of her baby until she stabilizes.

a plant-based, ethically manufactured bag that reminded her of our workshops. She came to the following week's workshop with her wallet and said she loved it. This indicated her level of interest was high and that she could grasp things once exposed to them.

At the end of the discussion point on sustainable consumption, one representative from each group explained that the labels would tell them more about the working ethics and mentality of the companies where they were produced, as well as whether they were beneficial to the environment, human health, etc.

The participants were shocked when they saw the slide showing how much waste a Hong Konger produces each day. Then, we asked about the importance of waste management and what might happen if they did not consider it. They responded by saying it would influence their health because waste could contaminate water, plants, and human beings. This demonstrates their understanding of the connectedness and impact of unsustainable living had increased to some extent

- **A mix of anthropocentric and eco-centric values**

Participants put forward eco-centric and anthropocentric values for preserving nature when asked why we should care about the environment. A good environment makes them feel good. They also acknowledged that unsustainable living would adversely impact nature and their lives.

> (I feel) very good. Fresh air, good smell. Good. Everything. We can walk. You can smell like a florist. You can smell like a tree. I like that. You can exhale and inhale properly. Ms. Daisy (Interview)

Image 5.3—They see that they bring adverse effects on their health and living environment. They agree that caring about the environment is a collective effort. They are willing to share relevant knowledge to people around them. They are able to understand what human actions' effects may have on the environment. However, they seem to incline towards a more anthropocentric position to care about the environment. One example quoted from the focus group is the use of pesticides in planting vegetables, which is then consumed by humans. Although they are aware of the use of chemicals on the environment, such awareness is not detached from considerations related to (it). Mr. Aaron (Observation notes)
When probed further about their interactions with animals and plants, the participants indicated that their children, in particular, enjoyed animals and plants.

"Florentin likes plants and animals. When asked if they were taking care of the plants we planted the first week, *"when he go shower, (he says) "teacher told me everyday (to water the plants)…When I saw the plants, too much water.. Now they're all gone, all died." Ms. Daisy (Interview)* (She is telling the story that he is making excuses to water the plant and play with the water).

Air quality was the most repeated environmental problem identified in relation with health issues. However, the connection was not always immediate.

> The air pollution, that's the number one problem I think and the waste management too. But the air pollution you know, every day we're breathing the polluted air. It might affect our

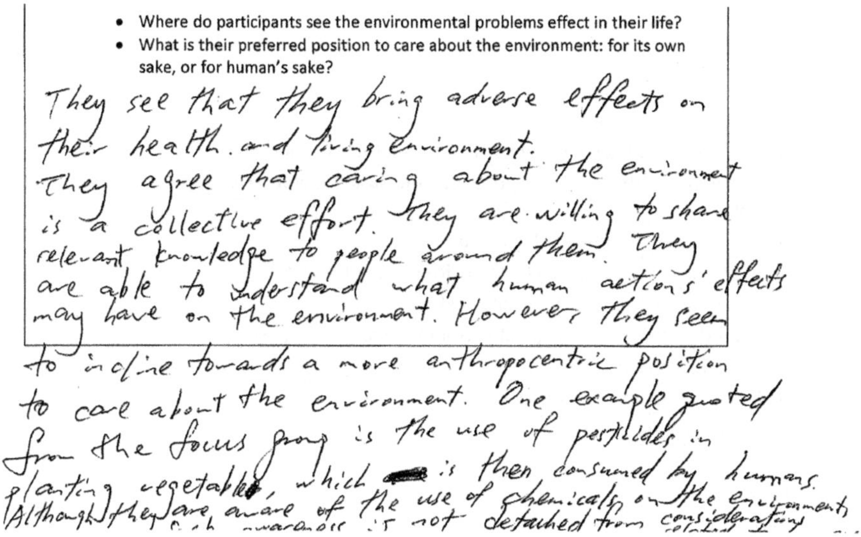

- Where do participants see the environmental problems effect in their life?
- What is their preferred position to care about the environment: for its own sake, or for human's sake?

They see that they bring adverse effects on their health and living environment. They agree that caring about the environment is a collective effort. They are willing to share relevant knowledge to people around them. They are able to understand what human actions' effects may have on the environment. However, they seem to incline towards a more anthropocentric position to care about the environment. One example quoted from the focus group is the use of pesticides in planting vegetables, which is then consumed by humans. Although they are aware of the use of chemicals on the environment, such awareness is not detached from considerations related to …

Image 5.3 Environmental values of RAS—Observation notes by the NGO representative

> lungs someday or when we grow old. So I'm scared also with my daughter because she's been living here for and been born here for so long. Ms. Rose (Interview)

- **Need for more exposure and education**

Participants said that education is essential for learning and can improve their participation in greening efforts, such as waste management and sustainable consumption. They believe education and investment in research are effective ways of tackling environmental problems.

The survey results indicate (Table 5.4) that education and investment in research are considered effective ways of tackling environmental problems.

- **The appropriateness of ESD**

As Mr. Aaron's observation notes indicate, the participants found the learning materials easy to understand and the pace easy to follow. They also enjoyed having their children with them and learning with other RAS women.

Among the participants, those with a higher education background seemed more interested in reading the materials and were more engaged during group discussions (Reflection memo—Week 3- Waste Management). Ms. Daisy mentioned her head was filled with daily stress, so she did not read. Some indicated they only skimmed through the materials the night before the workshop (Reflection memo—Focus group interview). However, if the materials were translated into their mother tongue, they could be more motivated to read (Observation notes—Mr. Aaron; Reflection memo—focus group interview).

The daily application of the sustainability issues provided on the concept information sheet, the use of case studies from the Hong Kong context for waste

Table 5.4 Survey result Q.7 (post-workshop)

Field	Count
Changing the way we consume	6
Changing the way we produce and trade	5
Investing in research and development to find technological solutions	5
Providing more information and education, e.g. on waste separation, energy consumption	8
Introducing stricter environmental legislation	7
Introducing heavier fines for breaches of environmental legislation	5
Encouraging businesses to engage in sustainable activities	5
Making the food system more sustainable from production to consumption	5
Using financial incentives (e.g. tax breaks, subsidies) to protect the environment	1
Ensuring better enforcement of legislation	1
Making the banking and insurance systems more environmentally friendly.	1
None	0
Don't know	0
Other (please specify)	0

Q.7—In your opinion, which of the following would be the most effective ways of tackling environmental problems in Hong Kong?

management and sustainable consumption, and the use of parts of a video of Mr. Indigo's interview was deemed helpful for three reasons:

(i) they made the topics more relatable to refugees' lives,
(ii) they stimulated their interest in SD issues and group work and
(iii) they encouraged them to reflect on alternative solutions, i.e., SE idea (Reflection memo—Weeks 2, 3, and 5).

Participants felt free to voice their opinions and become engaged during group discussions and activities (Student observation notes—Week III—Waste Management). Even though it was difficult for them to relate environmental issues due to their experience of daily stress, the more they were probed, the more they reflected on their actions, their impact on the environment, and their health, living, etc. The reflection points and discussions revealed that their survey responses and actions do not necessarily match (Reflection memo—Week 3—Waste Management).

These findings indicate that the aspects the refugees indicated they were content with, such as group work, group discussion, adaptability to household use, relatability, and references to local issues, are per what ESD suggests utilizing while teaching and learning.

- **The adaptability of skills to household use**

The participants found they could apply the skills they learned in their household, especially thanks to the hands-on activities. For example, Ms. Poppy said she did not know that recycling soap and making candles was easy, but now she can do it at home. Ms. Daisy also approached us after the workshop and asked where we got the candle-making soy wax.

> One of the ways I was pleasantly surprised was how some mothers, when wrapping their products in cellophane, seemed concerned about whether or not it was recyclable as a plastic product, showing quite an application of their critical thinking and cognitive competencies. Additionally, they seemed to develop an interest and understanding regarding the upcycling of certain household products, but whether they would consider taking these steps in their personal lives is debatable. (Reflection notes by the student helper—Week 3—Waste Management).

Table 5.5 illustrates the areas influenced by participation in workshops at various stages, including environmental knowledge, behavior, and attitude. We also inferred the potential impact of participating in social enterprises that were covered as part of the workshops. The findings are shared in sect. 2.

Table 5.5 Summary of findings of workshop evaluation

Workshop stage	Learning experience	
Before	• Environment not a priority, but willing to learn new skills and knowledge • Low level of awareness and misinformation about waste • Pro-environmental action without conscience	*"Their motivation was to learn skills and knowledge that are new for them" Mr. Aaron*
During	• Limited understanding of the complexity and connectedness of SD issues • A sense of community • Variety of (business) skills, cultural perspectives, and experimenting • SE missions influenced by background and workshops • More time to conceptualize SE	*"... participants formed good relationships and helped one another during the workshops. Ms. Jasmine, a south Asian asylum seeker...indicated higher communication (interpersonal) skills than others in this workshop. However, this behavior does not necessarily appear in the greening context. So, we noted that it could be channeled into greening via education" (reflection memo—Week 2—Sustainable consumption).*
After	• Increased awareness and willingness to learn about sustainable development • A mix of the anthropocentric and eco-centric values • Need for more exposure and education • The adaptability of skills for household use • Consideration of being a part of the solution through SE	*"They agree that caring about the environment is a collective effort. They are willing to share relevant knowledge to people around them." Mr. Aaron*

5.2 Section 2: Findings Related to Empowerment and Potential integration of RAS

This part relates to findings on empowerment and potential local integration of refugee women through social entrepreneurship training (developed as part of generic green skills training).

5.2.1 *Societal Level*

- **Society's low level of expectations and an unknown area for NGOs**

Society's expectation of refugees is reported to be low. Furthermore, legal aspects concerning setting up an SE are unclear from an NGO's perspective. It requires a local person for registry purposes. Despite being an ambiguous area, setting up an SE can improve communication between the RAS community and Hong Kong society and alter society's attitudes towards RAS, according to Mr. Indigo. According to Mr. Aaron, setting up an SE can become a bridge in terms of helping RAS to integrate into society.

As indicated by Mr. Indigo, an SE founder, setting up an SE in Hong Kong involves technical and legal aspects. The technical part concerns the value of the product and connecting with the right people, such as team members or investors. The legal aspect refers to registration. Legally, a passport is needed, but technically, there is a large volume of documentation.

> I guess people don't expect much from the refugees, which is sad. Or, you know, people probably think oh it must be a refugee. They must have gone through so much tough things. Why should I say such a tough feedback? You know, there is no growth without honest feedback, and in one of the things I always tell people like never buy my product out of pity, you know, get it because you think it's going to benefit you. That's it. Mr. Indigo (Interview)

NGOs found it ambiguous and sometimes dangerous to suggest RAS set up their SE. However, they deem it a good idea for refugees to consider, as it forms a bridge between society and RAS.

> Although it is a great idea, it is very risky for organizations to advise refugees and asylum seekers to establish one, because they do not know themselves either. They need to understand the legality and dangers it may pose for the community…Nobody is sure of the full scale of refugees' setting up a social enterprise. Currently, the application in HK feels like a far stretch; however, it is a good start for refugees to see the value and embrace what sort of value a SE might bring in their lives and maybe eventually consider establishing one day as well. Mr. Aaron (Interview)

5.2.2 *Personal Level*

- **Limited Use of Agency and Self-realization in HK**

Setting up an SE idea is very appealing, and our activities seemed to generate ideas for SEs. However, instead of considering it as an option in Hong Kong, the participants considered it for their home country.

We probed why, and RAS considered setting up an SE challenging primarily due to financial and legal reasons in Hong Kong, limiting their use of agency for HK (Fig. 5.1).

Most participants were attending such workshops for the first time; therefore, consideration of the support they might need was limited. This indicates that the length of workshops should be longer than 2 weeks. The lack of opportunities to experience could be related to internalizing their limited rights in Hong Kong, resulting in low motivation, as well as a fear of falling into the illegal side of the workforce.

> I think they (HK government) cannot help any support with this (setting up a SE) because it's not their obligation. They're only helping us for the one that I mentioned,the limited support for the food and then for the health insurance. Ms. Peony (Interview)
>
> In Hong Kong, we are not allowed because we are not allowed to work, even (if) we would like to participate…It can influence me because maybe I can learn something that I can use it to make entry back to my country. Maybe I can also make it a business like the one, that we are doing the soap, even though it's only like leftover but if I collect mainly, then I can make another brand of the soap. Ms. Peony (Interview)
>
> (It is a) challenge to set up because if you don't know how… you really need to know more how to set up first and what to do… Where you going to get some help also to do the social program. Maybe if they (the government or society) have a program like that (like ours), then it's okay. I can learn more. Ms. Lilly (Interview)

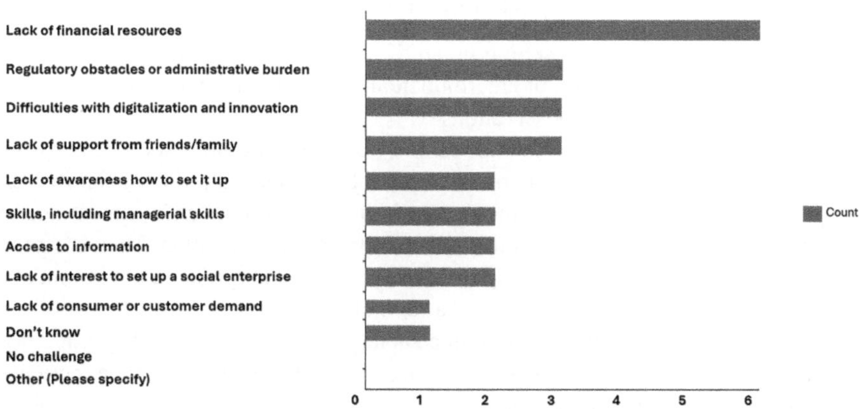

Fig. 5.1 Survey result Q.30 (post-workshop). Q30—If you could set up a social enterprise in Hong Kong, which of the following areas would pose the biggest challenges for you? You may choose maximum 3 answers.

To be honest, you know if you are in a place where you don't have much opportunity, right? If you're giving, you know, is refugee and then you're given the opportunity that you can create some product. You know the self realization that comes in with that is very very important… I developed Skills, project management skills. Everything in life is a project, and you know, even if being able to register for a company's project, being able to sharpen those project management skills, you know those things are lifelong skills. You know those are very lifelong skills that people do take for granted. Sometimes they don't think about it…And I can tell you it's much better to learn while doing than sitting in class…so I think it's a better, faster way to learn. Mr. Indigo (Interview)

5.2.3 Community Level

- **Sense of contributing to society and regaining dignity**

Women's motivation to set up or participate in SEs was found to be both anthropocentric[2] and eco-centric[3] reasons. As much as they would like to generate money and see it as a way to return home, they also believe they could contribute to society by managing their waste.

Maybe I contribute because I know how, where I can throw the good ones(wastes) and can recycle the other ones. and I can contribute also (to environmental problems) less and, uh, especially for the electricity, so it's not going to have a fire. Ms. Lilly (Interview)

We can apply it when, if ever we can go home or whatever. We can apply this, specially the candle, the soap making we can do that. We can generate money out of it. Ms. Rose (Interview)

Let's show you(to the government) that we (refugees) can do it without you (the government/society) first …we are hoping that once you realize that what we are not in this country on this city for handouts, free foods that we're here because we can give something else. We have had problems back home but we can give something else. Let them learn first. So that's where we are right now. We refuse to all they ask. Mr. Indigo (Interview)

- **Consideration of being a part of the solution through SE**

Participants are interested in setting up an SE or learning how to set one up. Their motivation seems to be a mix of environmental and economic concerns.

Mr. Aaron made similar observations.

Image 5.4—The participants showed interest in setting up a social enterprise. The motivation seems to be a mix of environmental, social and economic incentives. As much as they can appreciate the impact it may have on the environment and/or people around, income is also an important factor of consideration. They are positive in knowing more about the ways to set up a social enterprise for environmental preservation causes. Mr. Aaron (Observation notes)

The potential impact of participating in a social enterprise on societal, community, and personal levels for refugees is summarized in Table 5.6. The current experience indicates the need for empowerment.

---

[2] Focuses on the human benefits only.

[3] Focuses on environmental benefits only.

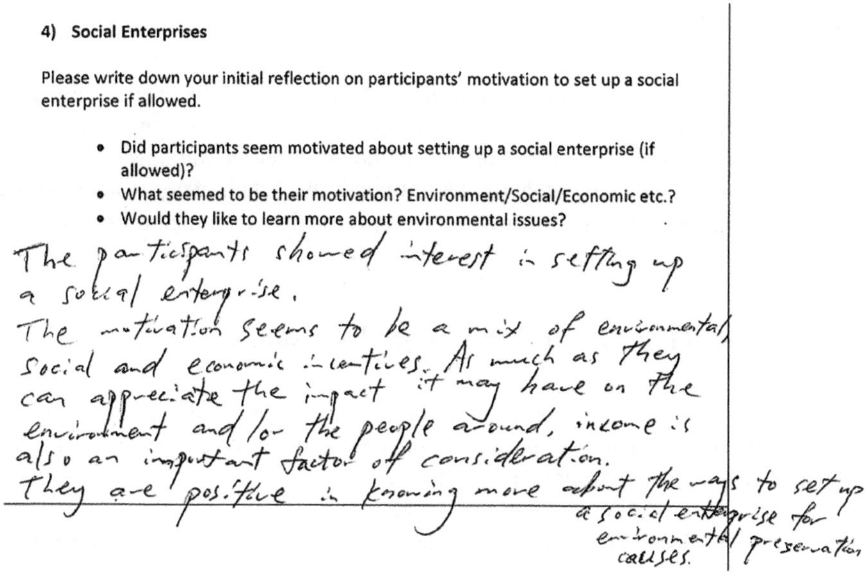

4) Social Enterprises

Please write down your initial reflection on participants' motivation to set up a social enterprise if allowed.

- Did participants seem motivated about setting up a social enterprise (if allowed)?
- What seemed to be their motivation? Environment/Social/Economic etc.?
- Would they like to learn more about environmental issues?

The participants showed interest in setting up a social enterprise.
The motivation seems to be a mix of environmental, social and economic incentives. As much as they can appreciate the impact it may have on the environment and/or the people around, income is also an important factor of consideration.
They are positive in knowing more about the ways to set up a social enterprise for environmental preservation causes.

Image 5.4 Motivation about SEs—Observation by the NGO representative

Summary of Findings

Table 5.6 Summary of findings about the potential impact of participating in SEs

Level of impact	Current experience	Possibility
Societal	**Society's low level of expectations from refugees and an unknown area for NGOs**	**Integration into society**
	"I guess people don't expect much from the refugees, which is sad." Mr. Indigo	
Personal	**Limited use of agency**	**Self-realization**
	"In Hong Kong, we are not allowed because we are not allowed to work, even (if) we would like to participate (in SEs)…it can influence me because maybe I can learn something that I can use it to make entry back to my country." Peony	
Community	**Sense of contributing to society**	**Regaining dignity**
	"Maybe I contribute (to society) because I know how, where I can throw the good ones (wastes) and can recycle the other ones. And I can contribute also (to environmental problems) less." Lilly	

5.3 Section 3: Discussion

This section discusses the findings regarding the workshop's effectiveness, based on the evaluative framework we outlined in Chap. 1, and the empowerment of refugee women through social entrepreneurship. Although intervention was designed to participants of all genders, after the recruitment, all of them were women, so we decided to explore literature further to focus specifically on women refugees.

5.3.1 About the Effectiveness of the Workshops

The effectiveness of the selected approach relates to whether and how generic green skills workshops, as a non-formal education, could influence RAS's perception and behavior toward the environment and sustainability.

The participants' first encounter with greening was the Green Refugee Day we organized. The NGO that collaborated and the participants provided positive feedback about this event, and they expressed an interest in the workshops in the following months. Therefore, Level 1—Reaction towards our evaluative framework was positive, and the result for the first level was successful.

The most significant result related to participation in generic green skills training was found to be in the development of cognitive competencies such as environmental awareness and a willingness to learn about SD issues and to think about things differently, such as turning environmental problems into an opportunity through SE, which indicates an achievement related to *Level 2—Learning*. Learning was achieved by simulating a real-world/home problem (which relates to a problem-oriented approach) in the classroom setting and generating solutions that could be applied in their contexts, meaning home (which relates to project-based learning). The problem (e.g., over-consumption) and its application in daily life and industry were identified in the concept information sheet, and a successful solution was introduced in the case study sheet. The activities involved discussion in groups and reflection on the topic. The workshops involved quantifying materials, while recycling sparked consideration of minimizing environmental impacts (e.g., using soy wax instead of regular wax). The workshops ended with hands-on activities involving small-scale solutions for their use and providing a holistic learning opportunity. Participants indicated they are now aware of the environmental impact of their actions and act more consciously. From this perspective, our workshop method aligns with studies that support the use of ESD pedagogy for teaching and learning about generic green skills (Pavlova and Chen, 2019; Shaping the Education of Tomorrow, 2012).

We tried to infer the applicability of the skills at home, *Level 3—Behavior*. In addition to a changed mentality toward sustainability, as a mindset and living style, participants indicated they adopt some of these activities at home, such as soap recycling. As a result, the workshops were influential in behavioral change, at least briefly when we interacted with participants. However, as suggested earlier, behavioral change may not be sustained unless it is supported by individual factors such as values and contextual supports. Therefore, the duration of this behavioral change is undetermined and more personal. Moreover, our study did not intend to explore the long-term effect on behavior.

Level 4—Results—The results show that the SE idea was new and inspiring to refugees and asylum seekers. Their motivation to establish a SE one day is related to financial and social circumstances. These motivations are not surprising, considering that Hong Kong is an expensive city to live in. Participants said our hands-on activities in the workshops, such as recycling soap, making toys/wallets out of scrap fabric, repurposing coffee grounds, etc., could be good ideas for a SE. However, the lack of financial resources and knowledge of setting up a SE was considered challenging.

As a result of participating in generic green skills workshops, participants indicated they paid more attention to environmental problems, their impact on them, and the attractiveness of SEs, which would be a solution-oriented approach to environmental problems. This is something they had not considered before. While reflecting on something they did not consider before and developing consciousness, participants indicated they felt they were at least contributing to the city and the environment by managing their waste. This feeling of contribution positively impacts their wellbeing and can be considered a resource gain. In support of this notion, international studies indicate that contributing to the benefit of others in society positively impacts one's wellbeing (Nelson, Layous, Cole, & Lyubomirsky, 2016; Nguyen, Kim, Daheim, & Neduvelil, 2020; Gruenewald, Liao, & Seeman, 2012). So, what had previously been 'background awareness' for them became an object of their consideration through workshops and so an object of their action and cognition (Freire, 1970, p.56). Enabling participants to think about things differently and to become a part of the solution concerning SD was empowering on a personal level because *"any education worthy of its name should always contribute to processes of subjectification that allow those being educated to become more autonomous and independent in their thinking and acting"* (Biesta, 2009, p.41). We cannot definitively ascertain the continuity of this impact and, hence, the behavioral change; however, based on the positive responses to our short-term training, we can argue that if RAS are exposed continuously to such opportunities (instead of legal problems each day), their observations and perceptions can change.

Pavlova (2022) suggests that personal relevancy, involvement, and self-efficacy mediate pro-environmental behavior in greening contexts (see Table 5.7). Based on this argument, our designed approach can be evaluated as a successful intervention because it addresses barriers through the recommended educational response. For example, participants recognized the relevance of waste management and consumption issues by applying these practices at home. They understood that they could contribute to the city's greening by managing their waste, which concerns their sense of self-efficacy. However, their involvement in the greening process, mainly through skills development and social enterprises (SEs), is limited by external constraints.

Freire (1970) argues, 'in problem-posing education people develop their power to perceive critically *the way they exist* in the world *with which* and *in which* they find themselves; they come to see the world not as a static reality, but as a reality in

Table 5.7 Mediating factors to bridge the behavioral gap

Barrier	Mediating Factors	Educational Response
1. Individual differences in the intensity or priority of values and attitudes	1. Personal relevancy and felt involvement towards green issues	• Prioritize weak anthropocentrism through values education and the overall context for program development. • Use real-life case studies and industrial examples of greening industries to initiate personal relevancy and involvement in the greening agenda to induce behavioral change
2. Lack of personal capabilities to engage in green behavior	2.1 self-efficacy 2.2 self-control	• Equip learners with green skills and self-efficacy to take green actions. • Ensure all students develop generic green skills. • Develop habits of green practices/operations during training programs
3. External constraints	3. Emphasis on the greening of industries and skills	• Prepare a workforce needed for addressing the structural changes of the green transition (e.g., policies on greening economies and skills, inclusion of green skills into training standards, developing programs that integrate these standards)

Source: Pavlova, 2022 (Based on Arbuthnott, 2009)

process, in transformation. Although the dialectical relations of women and men with the world exist independently of how these relations are perceived (or whether or not perceived at all), it is also true that the form of action they adopt is, to a large extent, a function of how they perceive themselves in the world' (p.56).

For our study, although environmental problems or their solutions may exist whether RAS perceives them or not, it is essential to recognize that they can play a role in this transformation and contribute their part. The workshops enabled RAS to reflect that they can contribute to society by managing their waste and, if allowed, through SEs based on their sharing. Therefore, the pedagogy of ESD can be considered empowering on both the personal and community levels, facilitating a dialogue and, eventually, fostering a culture of sustainability where everyone acts co-actively. This finding supports the benefits of ESD (e.g., Pavlova, 2019) and suggests that it may also be suitable for the refugee community, as it positively influences and involves RAS to reflect on their role in addressing environmental problems.

In summary, RAS's reactions towards training demonstrated that the participants considered any skills development to be meaningful. As indicated by the NGO representatives and the refugees, they enjoyed participating in the skills training and suggested continuing with the workshops, where they were actively engaged in doing something. However, generic green skills training proved to have added additional value due to its aims and potential area of influence concerning sustainable urban development because the participants indicated they had developed an awareness and conscience about the impact of their actions on the environment and others, especially during the workshop's discussions. Discussions allowed them to think reflexively. For this reason, allowing and/or easing adult refugees to access education and providing them a platform to improve themselves influences more than one person. In support of this perspective, numerous studies suggest that mothers' training and education have a positive influence on their children's educational outcomes (Magnuson, 2007; Borodkin, Orgal, & Martzini, 2022; Cui, Liu, & Zhao, 2019). Therefore, to ensure long-term behavioral change that supports greening, it is vital to involve the RAS community in the process and address external constraints, such as employment and access to training, so that they can contribute to sustainability. However, the international literature has highlighted the barriers that RAS women face in accessing labor.

5.3.2 About the Empowerment of Women Refugees Through Social Entrepreneurship

The global labor force participation rate for women is just over 50% compared to 80% for men. Women are less likely to work in formal employment and have fewer opportunities for business expansion or career progression (World Bank, 2022a, 2022b). The case is worse for refugees and asylum-seeking women. When women face various forms of gendered persecution, such as forced marriage, honor crimes, trafficking, sexual violence, or female genital mutilation, just because they are women (Quintero, Murray, Connelly, & Ballantyne, 2015), this can lead them to be asylees and seek safety elsewhere. Women and girls account for 49% of the people in refugee-like situations and other people in need of international protection (UNHCR, 2024). These women bring their skills, perspectives, and resilience shaped by their unique experiences to their new communities. Therefore, their participation in the workforce can be an opportunity to enrich society and economies, and they can help drive innovation and productivity across diverse industries. Moreover, Sustainable Development Goal (SDG) 5 calls for equal conditions, treatment, and opportunities for women and men. A report from European urban areas highlights the importance of refugee women's employment for several key reasons: integration into the host community, individual empowerment, and promoting respect for diversity and inclusion in urban development and economic growth (Spohn, Unzicker, & Vopel, 2018). Supporting this notion, Mulvey's (2013) findings suggest that refugees view employment as a means of overcoming social isolation, thereby helping them develop their language skills and social networks. However, RAS often face barriers in terms of access to employment due to language barriers, cultural stigmas, and lack of access to resources. To address this socioeconomic problem, studies and projects reported that social entrepreneurship and social cooperatives are effective solutions (Al-Hamad, Gare, Yasin, & Metersky, 2024; World Bank, 2022a, 2022b; Omran, Yousafzai, Al-Dajani, & Sheikh, 2021). Despite a substantial body of literature on the challenges faced by refugee women, a significant gap remains in research specifically focused on targeted educational and entrepreneurial interventions for this demographic (Khatib, Purwar, Shah, Vizcaino, & Castillo, 2024; Bikorimana & Nziku, 2023).

5.3.2.1 Barriers and Benefits of Inclusion in the Workforce

In this section, we highlight the challenges that refugee women face in seeking employment, as well as the benefits. Although some studies or statistics that are provided under each barrier are related to refugees as a group, as stated above, women experience the same barriers but to a higher extent.

Regarding refugee women's employment, it is reported to be a particular challenge to bring them to work. On average, the employment rate for refugee women in Europe is 45%. This is 17 percentage points lower than that of refugee men and six percentage points lower than that of other non-EU-born women (European Union, 2016).

The typical barriers to the labor market for refugee women are identified as below (Gettliffe & Khan, 2020):

- **Social/Cultural capital and norms**—*such as lack of networks and contacts, cultural barriers, inability to navigate in the system, lack of appropriate services to support transition to employment*

While the cultural acceptability of women working in the Middle East region may change, cultural norms and practices can still hinder women's employment. According to a survey of over 600 Jordanian and Syrian refugee women in Jordan, 43% reported that they perceived cultural, societal, familial, and religious pressures as an obstacle to women working (UN Women, 2018). In Germany and Kenya, the stories and experiences of refugee women demonstrated the economic empowerment and other benefits of refugee women's support networks. For example, in Germany, refugee women reported receiving support from German volunteers or sponsors, which, in turn, helps them with learning about German customs and norms, improving communication in German, accessing information about opportunities, and navigating the school system and housing (Gettliffe & Khan, 2020).

- **Workplace**—*discrimination, lack of knowledge of the host country's workplace culture, lack of specific assets, belongings, and savings*

Some studies reported the impact of fertility (motherhood) as a reason for not securing a position or returning to work for female refugees (Tervola, 2020; Quintero, Murray, Connelly, & Ballantyne, 2015). Gettliffe and Khan (2020) reported that a lack of effective childcare is a significant challenge in Germany in enabling women to attend training and other available options. In Australia, almost all of the participants in the study reported poor working conditions, exposure to toxic chemicals, as well as concerns about losing their jobs and immigration status. All these experiences and concerns negatively influence refugee women's health (Ziersch et al., 2023).

- **Education and skills**—*limited proficiency in the local language, low levels of qualification or lack of qualifications, lack of documentation for skills assessment, difficulties with recognition of skills and qualifications, and experience*

For instance, in Europe, the number of refugee women entering the labor market is reportedly low due to their home country's level of education and economic activity (European Union, 2016). On the other hand, Ruiz and Vargas-Silva (2013) found that women who are literate and have basic math skills are more likely to engage in outside employment.

- **Health**—*Psychological distress, poor physical health*

Coumans & Wark's (2024) study found that refugees face significant barriers to accessing and engaging with healthcare services, which impacts their integration into rural communities and increases the prevalence of psychosocial issues like feelings of loneliness, low self-esteem, a lack of autonomy, and a lack of empowerment over informed decision-making, especially for women, jobless men, and the elderly.

- **Housing and mobility**—*Limited affordable housing close to the workplace, lack of mobility (e.g. legal restrictions, transportation, safety)*

In Uganda, there is a 29.5% gap in productivity between male- and female-managed plots, not attributable to disparities in agricultural inputs. Ali et al. (2015) conclude that most of the gap can be explained by differences in men's and women's childcare responsibilities and transport access, which extension service providers often fail to consider when designing programs (Ali et al., 2015).

- **Legal status and labor code**—*temporary and uncertain residency, legal restrictions to enter the labor market*

A study in Germany reported that employers were more hesitant to offer jobs to refugees if there was uncertainty about their ability to stay in the country. This was a more significant challenge for those with "Tolerated Stay" or "Halted Deportation" status (Getliffe & Khan, 2020).

In Hong Kong, the barriers above seem similar to what our informants mentioned and what the literature revised in Chap. 2 suggests. These factors deny RAS' chances to integrate into society successfully. Local integration is a process with three interrelated dimensions: legal processes, such as the rights and freedoms accorded to these groups by the host state; economic processes, which are related to the penetration of the local labor market and attaining a degree of self-reliance for refugees; and social process which allows refugees to co-exist with other citizens in the host community without fear of discrimination or exploitation (Crisp, 2004). According to major refugee agencies such as the UNHCR, IOM, or NRC, a situation where these processes are not achieved is not durable (UNHCR, 2025a, 2025b; IOM, 2023; NRC, 2019). Durable solutions are about achieving a similar or better condition compared to the situation prior to displacement or compared to host communities (IOM, 2023).

5.3.2.2 Refugee Women Entrepreneurs

Despite these problems, many refugee women are still participating in the workforce. Ziersch et al. (2023) study highlights the impact of securing jobs on the well-being of refugee women in Australia. They found through in-depth interviews that once women are at work, their well-being is promoted due to feeling of contribution, developing a sense of identity, and extending social networks. In building their lives and contributing to their society, the entrepreneurial activities of refugee women are regarded as especially important. For instance, Edwige fled the Democratic Republic of Congo and resettled in Uganda. She learned to sew and runs a tailoring business in Uganda with agency support (Agbitor & Sunmonu, 2024). She said she wanted a vocational training school to teach people how to tailor their clothes. As a single mother of eight, she wants to teach single mothers, orphans, and school dropouts how to make money using their hands. Similarly, refugee women in Kenya and Niger spoke about their desire to improve their skills in entrepreneurship and business (Gettliffe & Khan, 2020).

Danielle, Zainabou, and Sheila are three women refugees who exemplify the incredible potential of 121,000 refugee women in Cameroon (Ngoh, 2023). Zainabou Abou fled the Central African Republic and found refuge in Cameroon. She learned to use beads to craft decorative items and fashion accessories. What started as a hobby soon became her sustenance, and she developed a thriving business that granted her financial independence. Sheila Marallah, another Central African refugee, took a different entrepreneurial path. She produces liquid soaps and bleach and sells them in her neighborhood. She stated that she was doing well financially before this journey, but now it was easier to get by. To help pay for her school fees, Danielle Siawali started selling snacks in secondary school. To find a more stable source of income, she learned how to make fascinators and hair accessories. Thanks to this activity, she can pay for her transportation and photocopies and meet her basic needs (Ngoh, 2023).

Fatima Khalif Abdi, a Somalian refugee in Ethiopia and mother of five, transformed her maize business with support from the Business Plan Competition, where she had to pitch her business plans to a panel of judges and industry experts. As a result of this competition, 50 exceptional businesses were awarded grant funding to launch or expand their ventures, along with ongoing coaching and mentorship to ensure their long-term success (ILO, 2024).

A study on women entrepreneurs in Glasgow found that 83% of refugee women entrepreneurs concentrated in restaurants, cooking classes, fashion boutiques, dry cleaning, and food supply sectors. Only 17% operated in accountancy services (Bikorimana & Nziku, 2023).

Kooy (2024) argues that financial security, independence, or previous small business experience in their home country are pull factors for refugee women who want to start a business. However, they have particular constraints on small business development, such as difficulties acquiring start-up capital, a lack of financial skills, limited access to affordable childcare, and fewer market-relevant support systems and networks than men. Expectations of family responsibilities, underpinned by religious restrictions and cultural norms, may add to these limitations. He suggests that even if women overcome these attitudinal barriers in their family and community towards women running businesses, they may still be considered responsible for childcare and home management, leading to conflict between work and family demands. Nevertheless, the incentive for many women to start small businesses springs from a desire for freedom from insecurity and to overcome traditional barriers related to language difficulties and financial and institutional constraints, such as ethnic and community solidarity.

Bikorimana and Nziku (2023) report that refugee women entrepreneurs wanted to give back to Scottish society by creating new jobs but were sometimes denied commercial loans. However, refugee women reported receiving a small amount of £1000 when taking out loans, repayable over 12 months, at a high interest rate of 26%. They also experienced racial discrimination, with their businesses being vandalized, and these went unreported to the authorities/police.

Al-Dajani (2022) argues that despite engaging in entrepreneurship successfully or otherwise, the refugee women remained marginalized and non-citizens in their host nation. This led him/her to question which comes first—entrepreneurship or empowerment. In other words, do the refugee women need to be "empowered" to delve into entrepreneurship, or do they engage in entrepreneurship first and, through its journey, become empowered? She found that if they are more willing to conform to the limitations and regulations imposed upon them by their families and/or husbands and other formal and informal institutions governing their economic participation, they are less likely to engage in entrepreneurship. So, she argues that a certain level of empowerment and elements of resilience are required for refugee women to behave and act entrepreneurially, and these can be elevated or downgraded through the entrepreneurial process.

5.3.2.3 Programs to Enhance Refugee Women's Skills

Lessons learned from international attempts to organize training indicate that programs should address more than one constraint, target host country vulnerable populations and refugees, include business training for potential entrepreneurs and address specific constraints women face by providing safe transportation and gender-based violence training. More evidence from evaluations of existing programs is needed to determine what works, particularly for promoting work opportunities for refugee women, especially in the post-COVID-19 contexts (Datta, Constant, Thawesaengskulthai, & Acheson-Field, 2020), which influenced the employability of refugees highly (World Bank, 2022a, 2022b).

To advance gender equality, foster social inclusion, and increase community resilience, the development of policies and programs for entrepreneurship and/or social entrepreneurship is recommended (Al-Hamad, Gare, Yasin, & Metersky, 2024; Hawkins, Abouali, & Al-Dajani 2018; Omran, Yousafzai, Al-Dajani, & Sheikh, 2021).

The dimension of social adds a social and/or environmental outcome rather than a financial one, and social entrepreneurship has three dimensions: sociality, innovation, and market orientation (Rivna & Gress, 2022). In the workshops of an initiative shared, workshops on subject lessons such as math or foreign languages, local language, and a buddy program that allows refugees to bond with locals and speak about their career desires and opportunities are shared as successful examples.

In another toolkit, trainees familiarize themselves with social entrepreneurship and the challenges of operating a cultural heritage craft enterprise (as the workshops focus on artisanship) with business development, and pitching skills are recommended (Hawkins, Abouali, & Al-Dajani, 2018). Given refugees' financial strain, offering complimentary refreshments and lunch and reimbursing them for travel costs was considered important (Buvinic & O'Donnell, 2016). Given refugees' limited access to the market, inviting guest speakers who can add value to the training sessions is another recommendation. A study in the UK found that having mentors of refugee backgrounds positively impacts refugee mentees' lives (Street, Ng, & Al-Dajani, 2022). Though the workshop for social entrepreneurship was spread over 5 days in Jordan, organizing it based on participants' availability and needs and dividing the participants into small working groups to increase harmony among members are recommended (Hawkins, Abouali, & Al-Dajani, 2018). Based on these, the skills development provided did not necessarily.

Our findings suggest that participation in or establishment of an SE serves as a means of self-actualization and engagement with the refugee community within society. However, the NGO representatives and participants highlighted the barriers preventing this community from contributing to greening and SUD through SEs. Barriers include society's low expectations of the RAS community and the ambiguity in legal and financial policies.

Participants are discouraged from pursuing social entrepreneurship due to personal and external barriers. For instance, they need more time to conceptualize SE concepts and connect to the environment, but regulations prevalent in the city discourage them. For this reason, they consider SE to be more feasible in the home country. This indicates that (i) there needs to be a more extended period of workshops focusing on SEs, and (ii) RAS have an expectation of returning to their home country, where they can apply the skills learned. Therefore, the perspective of refugees' investment in education is relevant to its applicability in real life.

Esser (2004) points out that investment in a particular resource in the destination country depends on the expected success and the cost. Our findings contribute to this by suggesting that the investment of time, energy, and money (e.g., transportation) in joining workshops may also relate to the potential success or applicability of the workshop in the country of origin. Being involved in social entrepreneurship training or such initiatives can increase the community's resilience and help them regain their dignity because they would contribute to their city and show they do not have to rely on governmental sources if given a chance. Pavlova (2022) and Arbuthnott (2009) argue that feeling involved is important to bridging the gap for behavioral change related to environmental problems. Moreover, social entrepreneurship is collective; it creates mutual support and the sharing of risks and responsibilities in the community. Therefore, it can empower the community (Vidovic, Peric, & Jozanc, 2015). It can also facilitate their local integration. Local integration can be observed at multiple levels: legal, economic, and social (Crisp, 2004). In terms of economic and social, our findings related to social entrepreneurship ideas showed that the RAS women's skills and perspectives (and resilience) are shaped by their unique experiences. If their participation in the economic sphere is ensured and cultivated through education and training, they can play a crucial role in driving (social) innovation in their new communities. However, to ensure this, legal changes must be made.

To address this structural change, the government must also recognize that RAS participation in SE at the community level has the potential to benefit society, the environment, and the economy and that this is a durable solution. Why should HK consider a durable solution? The RAS participants in our study, as well as many others, have reported being destitute in Hong Kong for a long time. Instead of contributing to the city by working, they try to survive with the minimum they receive from the government. As one participant noted, this leads them to explore illegal means of employment. This is harmful to themselves, the locals, and the city in the long term. As the European Union (2016) report shows, when the labor market is stable and the unemployment rate is low, the potential for refugees to thrive and contribute to their city is higher. In the case of Hong Kong, the unemployment rate has remained steady and low, at around 3% over the years, except during the COVID-19 period (Census and Statistics Department, 2025). The barriers mentioned are, in fact, the needs of refugees that are required for their well-being and empowerment. However, they are not accorded such freedom.

The impact of SEs on a country's economic development was identified in Europe. According to the European Commission (Summerfield, 2020), the social entrepreneurship sector employs approximately 40 million people, including those in RAS, and engages over 200 million volunteers globally—a sector that is growing. From a purely numerical perspective, significant social and financial benefits can be gained from reintegrating large groups of workers into the labor market. According to the OECD, social economy organizations (SEOs) account for around 6.5% of aggregate employment in the European Union (Summerfield, 2020).

A study in the Ugandan context found that refugees contribute positively to the host state's economy, based on the significant exchange volume between refugees and Ugandan nationals, as well as the creation of employment opportunities for Ugandan nationals (Betts, Bloom, Kaplan, & Omata, 2014).

In our study, refugees' responses to non-formal training are positive, and they find the SE concept to be meaningful. The examples above from around the world also indicate that participation in skills training and social entrepreneurship benefits both the community and the city. Therefore, considering a platform through which RAS can give back to society helps both the host and refugee communities. Therefore, governments, societies, and refugees should collaborate to define what freedoms to provide and empower these groups.

Freedom is widely discussed in terms of positive and negative paradigms. As Isiah Berlin defines it, negative freedom is "to do or be what one wants to do or be without interference by other persons" (cited in Kaufman, 1962, p. 241), resulting in minimum freedom. One can be considered free if one does not interfere with another person's freedom. However, the definition of freedom becomes significantly limited when we ask why we would like to do something, which leads us to positive freedom (Mamlok, 2016).

Broadly speaking, while negative freedom refers to '***freedom from*** interference, coercion or restraint,' positive freedom refers to '***freedom to*** act, to be as one will and self-determination' (Heyman, 1992). For human beings to realize their full potential, obstacles need to be removed, such as access to education. Mamlok (2016) states that if we want people to have rational and free choices, society needs to provide basic knowledge to make reasonable decisions. The conventional or classic approach argues that tradition allows people to live a meaningful life, whereas modernity has been seen to offer people freedom. It is problematic to distinguish between because, in classical thought, one is offered a good life in exchange for precluding from certain liberties, which is considered to safeguard the public interest (Heyman, 1992).

On the other hand, modern thought considers freedom to be the result of one's own ability to decide (Kalin, 2021); (Fletcher, Romero, Talbot, Warburton, & Whiston, 2020). From this perspective, how can HKSARG support refugees in having a meaningful life if they are not granted certain liberties such as employment, access to education, residence permits, etc.? Based on the participants' and NGO representatives' statements and a review of the literature, we understand that the government's position may stem from the notion that the provision of such freedoms would create a magnet effect for other refugees (pull factors), which might disadvantage the city's citizens.

Moreover, HKSARG is not obligated to provide employment or residency permits, as it is not a signatory of the UN Convention on Refugees. Nevertheless, the government fulfills its protection obligation through the non-refoulement principle of international law. From the Hong Kong government's perspective, it might be taking a positive constitutional approach. However, refugees' experiences and sharing strongly suggest the government is restraining them and *limiting their freedom to act and use their agency in ways that could benefit society*. Therefore, from the RAS perspective, the government's approach can be considered negative because the system does not allow them to exert their agency fully. However, refugees leave their home countries because of the existing barriers to their positive freedom and expect to receive it in the country of arrival; however, in Hong Kong, they do not. As a result, this clash between expectation and reality contributes to their ill-being.

Crisp (2004) argues that many refugees will likely stay in the countries they have fled to for a long time because the conflicts in their home countries are not ending soon. Just keeping these refugees in camps or without allowing them to move around and get jobs and education is harmful to the host society because it stops refugees from reaching their potential and helps the local economy. It can also lead to problems, especially with young refugee men getting involved in crime. If these refugees ever want to return home, they might not have the necessary skills or motivation when they cannot access jobs or education. Also, it may be true that the international community has not helped much with helping refugees settle in local areas; it is not correct to say that refugees cannot live peacefully and successfully in their new countries. Many refugees in developing places, especially in Africa, manage to take care of themselves without help from outside and live with local people. When looking for long-term solutions, it is important to consider these refugees' often unstable legal situation. The durable solutions states should consider are voluntary repatriation to the country of origin when it is safe, resettlement to a third country, and local integration (UNHCR, 2025a).

However, in any refugee situation, states should consider these options simultaneously with the circumstances and characteristics of different individuals and groups within the refugee population (Crisp, 2004). Refugees with the desire, the potential, or the need to become locally integrated should be enabled using appropriate legal and assistance measures. Refugees who have already attained a good degree of social and economic integration should be given secure legal status and residence rights, including the opportunity to become naturalized citizens of the asylum country. Even when local integration does not appear viable for many refugees, self-reliance and local settlement should be vigorously pursued as an interim measure. Such a strategy runs counter to the prevailing climate in many parts of the world and, if it is to be realized, will require the political will of host countries, the financial resources of donor states, and the expertise of development organizations (Crisp, 2004, pg. 7).

The HKSARG's attitude of "this is what we can offer and be content with it" towards RAS is not a remedy and is somewhat meaningless, and "meaninglessness cannot be considered a meaningful living" (Kalin, 2021). The Hong Kong SAR Government overlooks the capital and skills refugees bring with them and therefore limits their effective freedom, hence their capability. As discussed in an earlier chapter, media representations and society's impressions consider this community lazy and incompetent because they expect very little of them. Some even consider them to be abusers of the system. An empowering approach acknowledges that social, political, economic, educational, or cultural factors influence one's capacity to perform and offer positive freedom to act and do. Although adult refugees have limited access to employment and skills training in Hong Kong, they value education and are willing to learn new skills and knowledge. People's continuous relationship with their environment and society implies a need to establish the conditions in which an individual's capabilities might be realized either through a range of institutions or through individuals in the community by shaping social arrangements so they might better address their particular conditions. For this to occur, everyone must recognize that each person can contribute to the development of their communities through equitable negotiation.

5.4 Summary

We acknowledge that refugees bring resources from their country of origin, such as (environmental) knowledge, practice, and culture, or gain new resources in their new settlements. However, they are challenged by multiple factors in their new destinations, and Hong Kong is no exception. Hong Kong's strict rules about not recognizing refugees and treating unemployment negatively influence the community's well-being. However, including refugees and asylum seekers in greening and environmental education is necessary for creating a common culture and meaningful life because environmental problems affect all of us, regardless of our race, religion, country, or gender. Domains that influence wellbeing, such as supportive personal relationships, robust and inclusive communities, good health, a healthy and attractive environment, and values of democracy and social justice, can be nurtured through skills training (NESC, 2009).

The findings of our study indicate that RAS consider any skills development meaningful. As indicated by the NGO representatives and the refugees, they enjoyed participating in skills training and suggested continuing with the skills training workshops, where they were 'doing' something. Based on Sen's capability approach (1999), having the opportunity to do something that makes people happy (utility) is related to effective freedom, and in turn, increased effective freedom improves well-being. However, generic green skills training proved to have added potential due to its aims and potential area of influence concerning sustainable urban development. The participants indicated they had developed an awareness and conscience about the impact of their actions on the environment and others, especially during the workshop discussions. Discussions allowed them to think reflexively. For this reason, allowing and/or easing adult refugees to access education and providing them a platform to improve themselves influences more than one person. In support of this perspective, numerous studies suggest that mothers' training and education positively influence their children's education (Magnuson, 2007; Borodkin, Orgal, & Martzini, 2022; Cui, Liu, & Zhao, 2019). However, to ensure long-term behavioral change that supports greening, having the RAS community involved in the process, and addressing external constraints such as generic green skills training is vital. States' willingness to create durable solutions is important to safeguard refugees' rights and well-being and promote a welcoming attitude among host communities, thereby promoting socially just and inclusive development in urban areas.

Spinoza says, *the highest activity a human being can attain is learning for understanding because to understand is to be free*.' Although RAS indicated a lack of knowledge concerning sustainability at the beginning of workshops, their understanding and practices improved and positively changed due to participating in the workshops. They indicated they valued learning new skills through generic green skills workshops, and they could be applicable at home. Though the environment is not a priority for them, the workshops provided them with a valuable option to consider, i.e., setting up an SE in the future. Therefore, it can be regarded that participation in workshops augmented their capability.

Based on our findings and discussion, the next chapter presents our recommendations for an inclusive and sustainable approach.

References

Agbitor, K., & Sunmonu, R. (2024, June 20). *Refugee women build new lives and thriving businesses in East Africa*. Accion: https://www.accion.org/refugee-women-build-new-lives-and-thriving-businesses-in-east-africa/

Al-Dajani, H. (2022). Refugee women's entrepreneurship: Where from and where to next? *International Journal of Gender and Entrepreneurship, 14*(4), 489–498. https://doi.org/10.1108/IJGE-06-2022-0090

Al-Hamad, A., Gare, C., Yasin, M. Y., & Metersky, K. (2024). Refugee women's entrepreneurship: A concept analysis. *International Journal of Social Science Research and Review, 7*(4), 324–340. https://doi.org/10.47814/ijssrr.v7i4.2031

Ali, D., Bowen, D., Deininger, K., & Duponchel, M. (2015). *Investigating the gender gap in agricultural productivity: Evidence from Uganda* (7262; Policy Research Working Paper). Retrieved from https://openknowledge.worldbank.org/server/api/core/bitstreams/1eb59781-dabe-58f3-a34d-c972d8e75941/content

Arbuthnott, K. D. (2009). Education for sustainanle development beyond attitude change. *International Journal of Sustainability in Higher Education, 10*(2), 152–163.

Betts, A., Bloom, L., Kaplan, J., & Omata, N. (2014). *Refugee economics: Rethingking the popu;ar assumptions*. University of Oxford. Retrieved from https://www.rsc.ox.ac.uk/files/files-1/refugee-economies-2014.pdf

Biesta, G. (2009). Good education in an age of measurement: On the need to reconnect with the question of purpose in education. *Educational Assessment, Evaluation and Accountability, 21*, 33–46. https://doi.org/10.1007/s11092-008-9064-9

Bikorimana, C., & Nziku, D. M. (2023). Gender and entreprenurship amongst refugee women entreprenurs in Glasgow, Scotland. *Studia Prawno-Ekonomiczne, 127*, 99–139. https://doi.org/10.26485/SPE/2023/127/7

Borodkin, K., Orgal, R., & Martzini, N. (2022). The role of mother's education and child gender for children's vocabulary and math skills in the transition from early childhood education and care to first grade in Norway. *International Journal of Bilingual Education and Bilingualism*, 1–12. Retrieved from https://www.tandfonline.com/doi/citedby/10.1080/1350293X.2022.2055101?scroll=top&needAccess=true

Buvinic, M., & O'Donnell, M. (2016, November). *Revisiting what works: Women, economic empowerment and smart design*. Center for Global Development: https://www.cgdev.org/sites/default/files/CGD-Roadmap-Update-2016.pdf

Census and Statistics Department (2025, January 20). *Table 210-06101: Statistics on labour force, employment, unemployment and underemployment*. Labour Force, Employment and Unemployment. HKSARG. Retrieved from https://www.censtatd.gov.hk/en/web_table.html?id=210-06101#

Coumans, J., & Wark, S. (2024). A scoping review on the barriers to and facilitators of health services utilisation related to refugee settlement in regional or rural areas of the host country. *BMC Public Health, 17*(4). https://doi.org/10.1186/s12889-024-17694-9

Crisp, J. (2004). *The local integration and local settlement of refugees: A conceptual and historical analysis*. UNHCR. Retrieved from https://www.unhcr.org/sites/default/files/legacy-pdf/407d3b762.pdf

Cui, Y., Liu, H., & Zhao, L. (2019). Mother's education and child development: Evidence from the compulsory school reform in China. *Journal of Comparative Economics, 47*(3), 669–692. https://doi.org/10.1016/j.jce.2019.04.001

Datta, N., Constant, L., Thawesaengskulthai, N., & Acheson-Field, H. (2020). *Addressing employment obstacles for young Syrian refugee women.* World Bank. Retrieved from http://documents.worldbank.org/curated/en/315871591934181461/Addressing-Employment-Obstacles-for-Young-Syrian-Refugee-Women

European Union. (2016). How are refugees faring on the labour market in Europe? *European Commission and the OECD.* https://doi.org/10.2767/350756

Esser, H. (2004). Does the "new" immigration require a "new" theory of intergenerational integration? *The International Migration Review, 38*(3), 1126–1159. Retrieved from https://www.jstor.org/stable/27645428

Fletcher, R., Romero, P., Talbot, M., Warburton, N., & Whiston, A. (2020). *Philosophy: A visula encyclopedia.* DK Publishing.

Freire, P. (1970). *Pedagogy of the oppressed.* (M. B. Ramos, Trans.). Penguin Books.

Gettliffe, E., & Khan, S. (2020). *Scaling economic opportunities for refugee women: Understanding and overcoming obstacles to women's economic empowerment in Germany, Niger and Kenya.* International Rescue Comittee. Retrieved from https://www.rescue.org/sites/default/files/2020-10/IRC_WomensEcon_Cross-Country_Overview_vf.pdf

Gruenewald, T. L., Liao, D. H., & Seeman, T. E. (2012). Contributing to others, contributing to oneself: Perceptions of generativity and health in later life. *The Journals of Gerontology: Series B, 67*(6), 660–665. https://doi.org/10.1093/geronb/gbs034

Hawkins, M., Abouali, D., & Al-Dajani, H. (2018). *Social Enterprise training toolkit for artisans.* Retrieved from https://haqqi.s3.eu-north-1.amazonaws.com/2018-10/Artisans_TOOLKIT%20English%20Finalest.pdf

Heyman, S. J. (1992). Positive and negative liberty. *Chicago-Kent Colloege of Law Review, 68*(81), 81–90. Retrieved from https://scholarship.kentlaw.iit.edu/cgi/viewcontent.cgi?referer=https://scholar.google.com/&httpsredir=1&article=1306&context=fac_schol

ILO. (2024, September 3). *From refugee to entrepreneur: ILO's mentorship transforms Fatima's small business.* Retrieved from ILO: https://www.ilo.org/resource/article/refugee-entrepreneur-ilos-mentorship-transforms-fatimas-small-business.

IOM. (2023, November). *Progress towards achieving durable solutions: IDP Resettelements in Sofala, Mozambique.* Retrieved from IOM: https://dtm.iom.int/sites/g/files/tmzbdl1461/files/reports/IOM%20Mozambique%20%E2%80%93%20Progress%20Towards%20Achieving%20Durable%20Solutions%2C%20Sofala%20%28November%202023%29.pdf

Kalin, I. (2021, March 23). Anlam ve Ozgurluk. *Kendi Gökkubbemiz.* (MyMecra, Interviewer) Retrieved September 14, 2021, from https://www.youtube.com/watch?v=poBUMQ5pTFU&list=PLns7Jmm7BK800_o8ih_ryGUd3nf1%2D%2DLKO

Kaufman, A. S. (1962, April). Professor Berlin on 'Negative Freedom'. *Mind*, 241–243.

Khatib, M., Purwar, T., Shah, R., Vizcaino, M., & Castillo, L. (2024, December). Empowerment and integration of refugee women: A transdisciplinary approach. *Humanities & Social Sciences Communications, 11*(1). https://doi.org/10.1057/s41599-024-03723-w

Kooy, J. v. (2024). *Refugee women as entrepreneurs in Australia.* Retrieved from Forced Migration Review: https://www.fmreview.org/vankooy/

Magnuson, K. (2007). Maternal education and children's academic achievement during middle childhood. *Developmental Psychology, 43*(6), 1497–1512. https://doi.org/10.1037/0012-1649.43.6.1497. PMID: 18020827.

Mamlok, D. (2016). Negative and positive freedom: Considering education and the digital world. *Philosophical Studies in Education, 47*, 88–97. Retrieved from https://files.eric.ed.gov/fulltext/EJ1113115.pdf

Mulvey, G. (2013). *In search of normality: Refugee integration in Scotland.* Scottish Refugee Council. Retrieved from https://www.scottishrefugeecouncil.org.uk/wp-content/uploads/2019/10/In-search-of-normality-Refugee-Integration-in-Scotland-PDF.pdf

NESC. (2009, October). *Well-being Matters: A Social Report for Ireland.* Retrieved from NESDO: http://files.nesc.ie/nesc_reports/en/NESC_119_vol_I_2009.pdf

Nelson, S. K., Layous, K., Cole, S. W., & Lyubomirsky, S. (2016). Do unto others or treat yourself? The effects of prosocial and self-focused behavior on psychological flourishing. *Emotion, 16*(6), 850–861. Retrieved from https://psycnet.apa.org/buy/2016-19956-001

Ngoh, H. (2023, October 10). *Showcasing the potential of refugee women to contribute to economic growth in Cameroon.* Retrieved from UNHCR Africa: https://www.unhcr.org/africa/news/stories/showcasing-potential-refugee-women-contribute-economic-growth-cameroon

Nguyen, N. P., Kim, S. Y., Daheim, J., & Neduvelil, A. (2020). Social contribution and psychological Well-being among midlife adults with chronic pain: A longitudinal approach. *PubMed, 32*(10), 1591–1601. https://doi.org/10.1177/0898264320947293

NRC. (2019, November). *Durable solutions for internally displaced people: In a nutshell.* Retrieved from Briefing Note: https://www.nrc.no/globalassets/pdf/briefing-notes/durable-solutions-for-internally-displaced-people/durable-solutions_briefer_eng.pdf

Omran, W., Yousafzai, S. Y., Al-Dajani, H., & Sheikh, S. (2021). *Acknowledging alternative perceptions of success among internally displaced female entrepreneurs: An intersectional perspective.* SSRN. https://papers.ssrn.com/sol3/papers.cfm?abstract_id=3867985

Pavlova, M. (2022). Unpacking greening and skills recognition in micro, small and medium enterprises. In M. Pavlova & M. Singh (Eds.), *Recognizing Green skills through non-formal learning* (pp. 3–22). Springer. https://doi.org/10.1007/978-981-19-2072-1_1

Pavlova, M. (2019, July 14). *Greening of the economy through partnerships: Issues and impacts on skills development.* (S. McGrath, M. Mulder, J. Papier, & R. Suart, Eds.) Handbook of Vocational Education and Training, 195-218. https://doi.org/10.1007/978-3-319-94532-3_116.

Pavlova, M., & Chen, C. S. (2019). Facilitating the development of students' generic green skills in TVET: An ESD pedagogical model. *The Online Journal for Technical and Vocational Education and Training in Asia, 12*, 1–20. Retrieved from http://tvet-online.asia/issue/12/pavlova-etal/

Quintero, M., Murray, N., Connelly, E., & Ballantyne, F. (2015). *Raising refugee women's voices: Exploring the impact of Scottish refugee Council'swork with the refugee women's strategy group 2011–2015.* Scottish Refugee Council. Retrieved from https://www.scottishrefugeecouncil.org.uk/wp-content/uploads/2019/10/Scottish-Refugee-Council_s-Women_s-Project-Impact-Report.pdf

Rivna, A., & Gress, M. (2022). Social entrepreneurship as a new business model. The case of refugee integration. *Economic Research-Ekonomska Istraživanja, 36*(1). https://doi.org/10.1080/1331677X.2022.2120044

Ruiz, I., & Vargas-Silva, C. (2013). The economics of forced migration. *The Journal of Development Studies, 49*(6), 772–784. https://doi.org/10.1080/00220388.2013.777707

Sen, A. (1999). *Development as freedom.* Oxford University Press.

Shaping the education of tomorrow. (2012). *Report on the UN decade of education for sustainable development.* UNESCO. https://unesdoc.unesco.org/ark:/48223/pf0000216472

Spohn, U., Unzicker, K., & Vopel, S. (2018). *Learning from the world: Good practices in navigating cultural diversity.* Bertelsmann Stiftung. Retrieved from https://www.bertelsmann-stiftung.de/fileadmin/files/Projekte/Vielfalt_Leben/Study_LV_Learning_from_the_World_Good_Practices_in_Navigating_Cultural_Diversity_2018.pdf

Street, C. P., Ng, P. Y., & Al-Dajani, H. (2022). Refugee women business mentors: New evidence for women's empowerment. *Sustainability, 14*(15). https://doi.org/10.3390/su14159154

Summerfield, R. (2020, May). *The impact of social entrepreneurship on economic growth.* Financier Worldwide Magazine. Retrieved 9 19, 2022, from https://www.financierworldwide.com/the-impact-of-social-entrepreneurship-on-economic-growth#.YyiATXYRW72

Tervola, J. (2020). Different selection processes, different outcomes? Comparing labor market integration of asylum refugees, resettled refugees and their reunited family members in Finland. *Comparative. Migration Studies, 8*(28). https://doi.org/10.1186/s40878-020-00183-4

UNHCR. (2024). *Global trends: Forced displacement in 2023*. United Nations High Commissioner for Refugees. Retrieved from https://www.unhcr.org/global-trends-report-2023

UNHCR. (2025a). *Durable solutions*. Retrieved from UNHCR: https://www.unhcr.org/sites/default/files/legacy-pdf/61a0fe634.pdf

UNHCR. (2025b). *Technical and vocational education and training*. Retrieved from What we do: https://www.unhcr.org/what-we-do/build-better-futures/education/higher-education-and-skills/technical-vocational-education-training

UN Women. (2018, August 10). *Empowerment through employment for Syrian refugee women in Jordan.* Retrieved from https://www.unwomen.org/en/news/stories/2018/8/featureempowerment-through-employment-for-syrian-refugee-women-in-jordan

Vidovic, D., Peric, J., & Jozanc, N. (2015). The role of social entrepreneurship in empowerment of women in rural areas of Croatia. *Economy of Eastern Croatia Yesterday, Today, Tommorow, 4*, 713–726. Retrieved from https://ideas.repec.org/a/osi/eecytt/v4y2015p713-726.html

World Bank. (2022a, June 20). *Creating jobs for refugee women in Türkiye through social cooperatives.* Retrieved from https://www.worldbank.org/en/news/feature/2022/06/17/creating-jobs-for-refugee-women-in-turkiye-through-social-cooperatives

World Bank. (2022b, January 10). *Female labor force participation.* Retrieved from Gender Data Portal: https://genderdata.worldbank.org/en/data-stories/flfp-data-story#:~:text=The%20global%20labor%20force%20participation,do%20work%2C%20they%20earn%20less

Ziersch, A., Miller, E., Walsh, M., Due, C., Callaghan, P., Howe, J., & Green, E. (2023). 'I really want to work for me to feel good myself ': Health impacts of employment experiences for women from refugee backgrounds in Australia. *SSM–Qualitative Research in Health, 3*. https://doi.org/10.1016/j.ssmqr.2022.100209

Chapter 6
The Way Forward: For an Inclusive and Sustainable Urban Development

Abstract This chapter summarizes the findings and concludes the book by describing the conditions for improving the well-being and empowerment of refugee and asylum-seeking women. It presents the contribution of research to the field and shares suggestions for future research based on the limitations we encountered during this study's fieldwork.

The study concluded that the refugees value participation in generic green skills and social entrepreneurship training, which contribute positively to their lives. However, they feel depowered due to policies and institutions. The study's contribution can be observed at theoretical and practical levels. The recommendations include easing access to employment and allowing access to non-formal training programs and formal programs such as technical and vocational education and training so that this community can contribute to sustainable urban development in Hong Kong.

Keywords Conditions to improve Well-being for refugees · Empowerment of refugees through training and employment · Theoretical and practical contribution of research

6.1 Summary of Research Findings

The research aimed to understand to what extent the development of generic green skills could improve the well-being of refugees and asylum seekers (RAS) and enable them to contribute to sustainable urban development (SUD) in Hong Kong. We also explored how social entrepreneurship, which is developed as part of generic green skills (GGS) training, could empower this community and contribute to their livelihoods from their perspective.

The research findings showed that refugees value personal and cultural resources (e.g., they want to stay in Hong Kong for their children's schooling and earn their living instead of being dependent). However, their access to what they value is

A. T. Saral, M. Pavlova, *Education for Refugees and Sustainable Green Urban Development*, Education in the Asia-Pacific Region: Issues, Concerns and Prospects 78, https://doi.org/10.1007/978-981-95-1680-3_6

limited: the government does not provide them with skills training because they are not allowed to work afterward, and attending higher education is complex, as the representative from a non-governmental organization (NGO) noted. Therefore, RAS in Hong Kong has restricted capability. This restricted capability leads them to find ways to survive, such as selling food cards to others in the RAS community or being involved in illegal work because they have few other options. These choices are not considered ethical by the RAS; rather, they are efforts for survival. Therefore, RAS feels incapacitated because they are not able to live lives they value, nor do they have options to choose from for a better life. Therefore, they need empowerment. As their challenges are often found to be related to external sources, empowerment is required mainly from external sources, such as policies and institutions. Since this was an educational study, our recommendations will mainly focus on that. However, we are referring to the political conditions identified by other studies as they are highly important to ensure empowerment through education.

6.2 Conditions to Improve Well-Being and Empowerment

Empowerment refers to an individual's or group's capacity to make informed choices and then transform those choices into effective actions and outcomes (Alsop et al., 2006). Therefore, empowerment is both a process and a product. However, certain factors influence this capacity, especially for refugees, who are displaced individuals. In addition to agency[1] as discussed in Chap. 2, an opportunity structure— a set of rules, including existing formal and informal institutions —shapes and constrains human interaction and behavior. Several factors interact regarding formal institutions, including public attitudes, government policies, and the portrayal of refugees in the media, as well as individual characteristics of citizens. They are difficult to isolate because they affect each other.

The formal institutions include governments and their policies. It is argued that social cohesion is one of the many factors influencing governments' decisions and is often used to justify restrictions and containment publicly. Governments are worried that more inclusive policies will lead to crowding out of host citizens, reduced social cohesion, and a backlash against the politicians who facilitated refugees' access (Aksoy et al., 2024). A similar approach is observed in the context of Hong Kong, and it was examined by using the notion of de-politicizing strategy in Chap. 4.

Loper (2010) suggests that the UN Refugee Convention should be signed by the Hong Kong Special Administrative Region Government (HKSARG) because the standards would consolidate the grounds for realizing refugee rights and establish a basis for legal representation in domestic law, combined with existing human rights provisions. This would lead to recognition and the right to employment, which

[1] Agency is related to a range of resources an individual or group has, such as social, personal, and material resources, and choosing civic, economic, or social actions (Alsop et al., 2006).

appear to be one of the significant reasons that negatively impact RAS's well-being in the city.

Globally, marginalized groups without political or economic power, such as refugees and asylum seekers, are among the most likely to be affected by environmental degradation due to rapid urbanization (McGranahan et al., 2005). Considering that education (an opportunity structure) can equip participants/learners to use agency and learn behavior, allowing this community to access training and education, especially generic green skills, could transform their understanding of urban and environmental issues and integrate them into the city's efforts for greening.

In urban areas, being engaged in nature and the environment has been proven to improve one's health and well-being (Bratman et al., 2015). So, the provision of generic green skills training and social entrepreneurship can be of value to the refugee and asylum-seeking community and can benefit the city's urban development for several reasons:

(a) It would give them an option to choose from, hence improving their effective freedom because when the value of functioning available is improved, so is the value of capabilities and effective freedom (Sen, 1999).
(b) it can be a step forward for people to realize the importance of nature on their wellbeing and act environmentally friendly. This could replace refugees and asylum seekers'lack of self-esteem because they may gain dignity and feel valuable for contributing to the community and the city. The Victorian Association for Environmental Education (2006) also confirms that people with a strong sense of belonging to their community tend to develop a higher sense of responsibility towards their living environment. Therefore, refugees would be more willing to act to protect the urban ecology (cited in Lee & Chan, 2009).
(c) When permitted, refugees may use their knowledge and skills to generate income and solutions for environmental problems and the community through social entrepreneurship. This could give them a purpose and make them feel happy.

Kennedy's (2011) description of the Education Bureaus principle of operation towards ethnic minorities in Hong Kong can also apply to the government's strategy for the RAS community's access to education: it is sufficient rather than equitable. This aspect resonates with the 'negative freedom' approach, as described in Chap. 5, whereby freedom is defined by actions free from interference and the coercion of others. So, the government's interference with issues related to this community only relates to jurisdictional necessities, which offer international protection for RAS but not the provision of continuing adult education. Education provides people with skills and knowledge to use their agency effectively and to enable them to 'do something' for their community and city. Therefore, access to education at all levels should be promoted.

For these reasons, our recommendations on improving the RAS women's capacity to contribute to sustainable urban development are based upon long-term thinking. We agree with Freire (1970) and Arbuthnott (2009) that for a transformative change to take place (in our case is the environmentally sensitive behavior adapted by all people in the society, including RAS), there must be a link between knowledge, understnding and actions, otherwise, positive change cannot be guaranteed to continue. Pavlova (2022) recommends the following for all groups of society to optimize greening processes:

a) they should be provided training and.

b) policies to include this community in such training and the workforce should be considered.

Below, we summarize the results of our study supported by the international examples that indicate the generalizability of our findings to a broader context.

6.2.1 Way Forward 1: Access to Non-formal Training and Formal TVET Education

Our study found that generic green skill development influences the well-being of refugees positively for several reasons: Participants learned new skills and knowledge they could apply at home, and they indicated they were happily participating in these workshops as a result of our intervention. Now that they are conscious of how their pro-environmental behavior influences the environment, they feel a sense of contribution to society by managing their waste, which positively contributes to their well-being. While developing this awareness and behavior, the use of Education for sustainable development (ESD) pedagogy during the workshops was effective because it enabled RAS to reflect on a relatable problem (waste management—used cooking oil), discuss their impact, and practice a solution that could be applicable in their daily lives (turning used cooking oil into candles or recycling soap). However, they still encounter problems systematically and daily. For example, their access to skills training, employment, or housing is restricted, primarily due to their lack of status. Therefore, although the SE idea was considered promising and helpful for regaining dignity, it is perceived as uncertain due to the lack of legal and financial resources. Despite the help from NGOs and other RAS community members, overcoming barriers requires intervention from the state because barriers result from policy and practices at the societal level.

At this point, it is important to acknowledge that entrepreneurship is not a simple solution for refugee women who face social and cultural transition issues related to long-term settlement in their host countries after experiencing significant multifaceted loss, precarity, and violence. Rather, it can be extremely dangerous, unstable, low-paying, and unprotected, particularly if it is done in the host countries' informal economy (Al-Dajani, 2022). Therefore, offering multiple lessons in non-formal settings that have some control over what is taught and who is teaching can be helpful at the initial stages. And certification of refugees' participation can be useful especially if they lost their education documents.

The Refugee Education Strategy for 2030 designed by the UN Refugee Agency (UNHCR) (2019) emphasizes the importance of girls and women having equitable opportunities for and access to education, work, community representation, and leadership. It recognizes the importance of TVET, allowing various opportunities and programs.

Our findings show a sense of hopelessness concerning accessing technical and vocational education and higher education among the RAS community. To pursue higher education is perceived as a highly complex and rare option for refugees and their children in Hong Kong, to the point where they question its purpose if they are not going to be able to pursue it further. However,

> a major function of education—of schools and other educational institutions—lies in the qualification of children, young people and adults. It lies in providing them with the knowledge, skills and understanding and often also with the dispositions and forms of judgement that allow them to "do something"—a "doing" which can range from the very specific (such as in the case of training for a particular job or profession, or training for a particular skill or technique) to the much more general (such as in the case of the introduction to modern culture or Western civilisation, the teaching of life skills, etcetera) (Biesta, 2009, p.39-40).

Internationally, many studies related to "skills training for refugees and asylum seekers" fall under technical and vocational education and training (TVET). Successful examples referred to in Chaps. 3 and 6 demonstrate that skills development allows refugees to contribute to their development and their new communities, improving their well-being. Another example is Switzerland. Although it has a highly decentralized provision of vocational skills training for refugees, the federal and cantonal authorities, professional training organizations, and employers developed a one-year pre-apprenticeship program for refugees. As a result, some 800 refugees were able to start a pre-apprenticeship program based in professional schools and firms (Aerne & Bonoli, 2021).

Similarly, a report from the US suggests assisting low-skilled immigrants in building a useful career ladder by linking vocational training and English instruction to promotions in certain sectors, including hospitality, dining, tourism, and healthcare. Vocational English as a Second Language (VESL) programs and bilingual job paths can give immigrants a foothold in growing areas that promote upward socioeconomic mobility. Organizations working with the community, vocational and technical education providers, and industry representatives should all be involved in designing such programs. This will supply skilled labor to local firms and in desired areas (Immigrant and Refugee Rights Coalition of Illinois, 2006).

In addition, TVET institutions can benefit both governments and RAS because TVET programs align with national labor market assessments and development goals (UN-Habitat, 2017). A government report indicates that TVET is considered inferior in Hong Kong (Task Force, 2020). Employers have expressed concern about the current job market mismatch, as many para-professional/technician positions requiring TVET training and knowledge remain unfilled (Task Force, 2020). However, TVET can potentially improve technical capacity and play a crucial role in greening economies (such as clean energy production), as Pavlova and Chen (2019). Why should refugees be intentionally the focus? Because what is required for moving from green economies to green societies requires creating a civil society engagement, including all groups of people, and developing 'consciences of contradictions' in societies (for example, a customer can buy a product if they s/he sees a 'green' tag on it without considering whether the company adopts an environmentally sensitive policy and/or practices) and problematizing the claims and strategies of the green growth (Wanner, 2015). Greening requires that all groups in society be engaged critically. Otherwise, green growth remains a myth, and capitalist societies and hegemony development continue (Wanner, 2015). Unless refugees, who are currently the only recipients of greening services in Hong Kong, develop a consciousness about it, the sustainability of efforts in the city may not be achieved. Therefore, people should adopt environmentally sensitive values, see themselves as contributing members of their society, focus on how they can change and challenge the problems, and innovate new ways to challenge ideas as a result of education for sustainable development. This is the emancipatory and transformative feature of education (Freire, 1970; Pavlova, 2012) because it allows them to become agents of change. Therefore, education is an empowerment tool and should be within RAS's reach. Also, allowing refugees to access TVET or non-formal training programs gives them a chance to participate and can increase their sense of choice/control in life and capability. This access and/or freedom to attend can improve well-being (Sen, 1999).

6.2.2 Way Forward 2: Easing Access to Employment and Workplaces through Establishing Social Enterprises

Refugees indicated their desire to work, yet they claimed that strict rules forced them to find illegal ways to earn money. This finding is similar to other study reports in the city (Vecchio, 2016; Mathews, 2018). In addition, it was mentioned that refugees' potential (lawful) employment is regarded negatively in the city, especially during COVID-19, when many locals were unemployed. This indicates a sense of competition in the public domain. Verme (2023) argues that areas that suddenly become home to refugees may experience resource strain and environmental degradation, which can negatively influence local development. In urban areas, the influx of refugees may lead to water and electricity shortages, causing higher costs for the refugees and the host community (for example, in housing), overcrowding of services such as health and education, increased traffic and pollution, and employment. These can lead to unemployment, especially among certain groups (such as the low-skilled and women), exploitation of workers, increased inflation rates, tension between the host community and refugees, and disparities in social and economic areas (Verme, 2023). These effects are observed in host communities in the short term.

Verkuyten (2021) argues that how public policies are framed (such as a potential threat to the host society or instead as making a valuable contribution and needing help) and how immigrants are depicted in the media and by politicians are important factors shaping society's attitudes toward newcomers (Verkuyten, 2021). The rules governing interactions between refugees and the host country, the structure of host economies, the characteristics of refugees, and the necessary skills in the host country all interact and impact the economic impacts of refugees (Taylora et al., 2016), even in the short term.

According to a new study, refugees from Ukraine have significantly contributed to the Polish economy. Almost a month after Ukranian refugees' arrival, the Polish government announced, "*A citizen of Ukraine can start working in Poland without a work permit. An employer only needs to remember to confirm this fact within 14 days on the website praca.gov.pl.*" (The Republic of Poland, 2022). The study estimates that Ukraine refugees contributed 0.7 and 1.1% to Poland's Gross Domestic Product (GDP) in 2023. Financial aid from the state made up just 5% of refugees' resources, whereas 80% was derived from work, according to estimates by Deloitte based on UNHCR data surveys among refugees (UNHCR Poland & Deloitte Polska, 2024).

Another study in the African context reveals that refugees, given the opportunity to interact with the economy around them, can create positive income spillovers for host-country households. Congolese refugees in Rwanda generate considerably more income than the cash aid they receive. However, spillovers are smaller when refugee aid is in the form of food instead of cash. Access to food supplies and other commodities, along with the cash to interact with the local economy, is critical to refugee welfare and refugees' potential to create benefits for the host country (Taylora et al., 2016). Although the study focused on camps, the argument is that refugees may positively influence markets if allowed not to rely solely on governmental sources. This corresponds to the external constraint in Pavlova's (2022) argument, as explained in Chap. 5, and requires policy to support the inclusion of RAS. Therefore, we will highlight some successful examples below that can apply to the context of Hong Kong and allow the government to develop more inclusive policies.

Peri's (2012) study shows that in states with many immigrants in the US, natives with lower education tend to take up communication-related positions, leaving manual-intensive jobs to immigrants. As a result, the author identified efficient task specialization among natives and immigrants, hence increased productivity gain (Peri, 2012). At this point, we would like to emphasize that we do not suggest confining refugees to be able to use low-level skills. In fact, (generic) green skills, as portrayed in our study, have added value because they contribute to the economy or society and the environment. Therefore, portraying RAS as a complement to the local market instead of a competitor to locals would be a positive approach.

Research shows that in 2015, refugees in the US earned a collective $77.2 billion in household income. They also contributed $20.9 billion in taxes, leaving them with $56.3 billion in disposable income, or spending power, to use at U.S. businesses (New American Economy, 2017).

A recent study from Syria shows that refugees with better conditions in exile are more likely to return (Beaman, Onder, & Onder, 2022). Similarly, growth at the destination is associated with a higher likelihood of return among Nepali migrants (Tiwari, 2021). The underlying assumption is that migrants prefer to consume money earned abroad in their country of origin (Tiwari, 2021). Constable's (2014) study in Hong Kong argues that a long waiting period allows claimants to earn money before returning to their home country. However, this is mostly illegal in Hong Kong, whereas in Jordan or Turkiye, there are a lot of Syrian refugees. Therefore, whether the goal is to return RAS to their origins or to integrate them into society, employment can be a constructive step. In this aspect, social entrepreneurship can be a positive and long-term avenue for refugees to participate. The study from the US shows that 13% of refugees were entrepreneurs compared to 11.5% of non-refugee immigrants and 9.0% of the U.S.-born population. The refugees' businesses also generated $4.6 billion in business income that year, highlighting the positive impact refugees' integration into employment can have (New American Economy, 2017).

Even if the legal platform existed, how would we ensure employers hire refugees? Turkiye's Transition to Formality Program (KIGEP) for Syrians[2] provides an example. This program aims to support the formal employment of both Syrian and Turkish workers, so they work under better conditions and to prevent unfair competition. Through incentives, the KIGEP covers the expenses of social security premium support payments for Turkish citizens and newly recruited Syrian workers for up to six months, as well as work permit expenses for Syrian refugees. To be eligible to apply for this incentive, enterprises must employ workers without interruption, pay their social security premiums, and be debt-free in the social security institution system (ILO Office for Turkey, n.d.). The European Union Trust Fund and the US Department of State, Bureau of Refugees and Migration co-founded the scheme. The scheme not only had a positive impact on productivity, especially for micro and small enterprises, but 41% of Turkish workers stated that after starting to work with Syrians, their opinion about working together changed for the better, indicating the impact on social cohesion (ILO Office for Turkey, n.d., p. 10). Similarly, Hong Kong could encourage enterprises to employ refugees legally and/or set criteria for eligibility.

6.2.3 Way Forward 3: Skills Profiling and Matching with City Needs

Besides policy, refugees' potential productivity is also related to their skills and market needs. Our findings showed that no skills-match services are conducted for RAS that can address the mismatch of available skills and industry. In Finland, creating a profile for refugees' skills and experience' is delegated to an NGO. Startup Refugees is a social innovation with over 1500 members, including companies, universities, government officials, and NGOs. This group has developed a database for

[2] Who are constitutionally not considered to be refugees; rather they are 'Foreigners under Temporary Protection'.

almost 6000 refugees with profiles that include their educational background, work experience, skills, needs, and motivation. They then match refugees with business experts and provide courses and mentorships to start businesses or employment (Startup Refugees, 2022). In this case, the legal platform has been established to employ refugees. Another European study reported that Germany, Italy, Romania, and Denmark have integration centers where refugees remain after arrival. During this time, a portfolio is collated of their competencies and skills, including an educational biography, records of professional experience, and formal and informal skills, which is intended to ease economic integration (Chadderton & Edmonds, 2014). As demonstrated by the literature, in Europe refugees are acknowledged legally and, despite challenges, attempts to integrate them into the economic system through formal education do exist. A similar approach—one that matches refugees' skills with areas in the industry can be applied for forward-looking purposes in Hong Kong.

The global trends regarding the demand for vocational education at the tertiary level are rapid technological advancements, digitalization, and the growing emphasis on sustainable development. and the sectors with increasing demand for higher vocational education include renewable energy and sustainability, ICT and digitalization, advanced manufacturing and robotics, healthcare and social services, creative industries and design agriculture, and sustainable development hospitality and tourism (UNHCR, 2025a, 2025b). Hong Kong Manpower projections also indicate that *information technology and information services, innovation and technology industries, cultural and creative industries, environmental industries, and testing and certification services* have good potential to drive economic development in Hong Kong. The manpower requirements of these selected industries are projected to grow at average annual growth rates ranging from 0.9% to 4.3% from 2017 to 2027, all faster than that of 0.3% for the whole economy (HKSARG, 2019). Thus, organizing training for RAS in those areas will benefit the economy and greening efforts of the HKSARG.

6.2.4 Way Forward 4: Noticing the Discourse

It is hard to dismantle social structures and forms of inequality because they are frequently maintained over long periods and involve deep structural values, which are influenced by one's employment status and job characteristics (Narayan & Petesch, 2007). It would be erroneous to expect the HKSARG to change its discourse quickly.

Understanding the role of the media in this discursive process is essential. However, discourse that focuses on complementary aspects of communities rather than competitive elements is a more inclusive approach that the media and the city's senior officials can take.

6.3 Contribution of Research

The contribution of this study can be considered at both the theoretical and practical levels. In terms of contribution to knowledge, the results helped to:

1. *Address a research gap*—This study addresses a gap in the literature concerning skills development for refugees and asylum seekers. It demonstrates that they can contribute to environmentally sustainable urban development by managing their waste following the provision and access to education. Refugees and asylum seekers have been overlooked as a group in the Hong Kong context regarding greening studies. The RAS in this study indicated they were happy attending these workshops and would like to continue.

In addition, the study combined several concepts, namely RAS, generic green skills, and SUD, a combination under-researched at the international level. Using the voices of refugees in our study, we found that these concepts complement each other. SE is an arena where all these concepts merge because refugees look at it holistically and recognize that the skills they learned can be applied to SE. Participation, or establishing an SE, can enable RAS to use its creativity and resources in ways that address environmental problems. Therefore, they can add value to their city. However, applicable legalities require amendment.

2. *Ensure Novelty*—As far as we can ascertain, this is the first study to consider generic green skills development to ameliorate RAS's lives in Hong Kong. The survey results and focus group interview indicated that refugees value the environment for both anthropocentric and eco-centric reasons. Their application of intentional pro-environmental behavior before the workshops represented their attempt to survive in this expensive city with the little subsidies they received. Therefore, it cannot be considered a sustainable attitude required for SUD. Based on interviews with NGO representatives, establishing or setting up an SE can build bridges between this community and society so they can learn more about each other and create an inclusive society. However, external barriers were identified to achieving this.
3. *Design the effective intervention*—The intervention assessed RAS awareness and practices concerning sustainability. Its effectiveness has been proven at different levels. The Green Refugee Day was the first occasion we met the refugees and introduced our study. It was a successful event, and we received positive feedback. As a result, the NGO was happy to collaborate with us again. Consequently, based on Kirkpatrick's evaluation model as indicated in Chap. 1, the first level—reaction, was positive. The second level—learning, involves improved skills, attitude, and knowledge. Participants came to hold sustainability as a mentality, purpose, or solution instead of considering it merely a catchphrase that everybody uses but does not comprehend or practice. The written

materials we used introduced each concept and how it could be applied daily. In addition, there were discussion points for each concept that refugees could reflect on and voice their or the group's opinion on, such as what made Health Riders a successful SE? (this was a case study presented in week 3). For the third level—behavior, which relates to applicability at home—refugees indicated they were happy to learn new skills and knowledge they could apply at home. This was possible due to the hands-on activities, such as turning used cooking oil into a scented candle. The fourth level—results is about showcasing the skills learned through SEs. This concept requires more time to become integrated into everyday thinking. However, the idea was inspiring to the community. Currently, it is not considered possible to execute in Hong Kong from the refugees' perspective, and NGOs find it somewhat untenable due to legal restraints. Therefore, establishing or participating in a SE is considered an option when and if the refugees return home.

4. *Proof of the effectiveness of ESD pedagogy and selected content*—Learning was achieved by simulating a real-world/home problem (which relates to a problem-oriented approach) in the classroom setting and generating solutions that could be applied in their contexts, meaning home (which relates to project-based learning). The problem (e.g., over-consumption) and its existence in daily life and industry were outlined in the concept information sheet, and a successful solution was introduced in the case study sheet. The activities involved group discussions and reflection on the topic. As a result, instead of adopting the passive learner and active teacher roles during the intervention, we adopted a student-centered approach and involved the RAS in their learning. The workshops ended with hands-on activities involving small-scale solutions for their subsequent use and providing a holistic learning opportunity. Participants used their creativity when producing materials during hands-on activities and created a business plan as a group. The participants left the workshops happy each week and said they wanted to continue. Therefore, the pedagogy and content were appropriate for this group. Since participants indicated being pleased about the results and contributing to society by managing their waste, they had positively impacted their well-being from Sen's (1999) perspective. This aspect matches our initial expectation of improving their wellbeing, so ESD can also be considered an effective approach.

At the practical level, our study contributed to the improved participants' wellbeing. Thus, at the personal level they considered generic green skills to be a valuable (cultural) resource gained. On the community level, RAS recognized the need for overall community efforts towards greening during the workshops. It was suggested by the SE founder, who is a refugee, that SEs can help refugees gain their dignity and contribute to the city meaningfully. However, more inclusive and emboldening rules and regulations are necessary to achieve this at the societal level. With altered rules and regulations, empowerment at all levels could be realized and maintained.

6.4 Limitations and Suggestions for Future Research

Below, we provide suggestions based on the limitations we encountered.

In this study, we reached only a small group of people within a particular context. Therefore, the findings cannot be generalized. However, as the analysis of the international literature suggests, very similar patterns can be identified internationally. In addition, no retention test was conducted in the months following the workshops to check whether the impact of the generic green skills workshops was retained. Therefore, our outcomes proved to have an immediate effect. In order to measure the long-term effect and create causality, conducting a study in which RAS are constantly and systematically exposed to this training and reminded of their relationship with nature, combined with a control group with opposite conditions, is suggested. Such a study would potentially ensure the continuance of the positive effect so it can be adapted for life, following the successful generic green skills training.

Another limitation relates to a lack of time and the small number of stakeholders. The workshops only ran for 4 weeks, and the study only involved refugees and NGOs. Therefore, the interpretation is limited to these participants. With this in mind, the study aimed to interpret concepts from the RAS'perspectives. Based on our interactions, participants enjoyed the workshops and expressed regret that they did not last longer. As a result, we recommend involving governmental bodies, NGOs, private sector investors, mentors, universities, and refugees in future studies. That way, the impact could be more comprehensive and large-scale, and the effects could be reflected at different levels, which would assist with advocacy and integration.

Social entrepreneurship workshops drew participants' attention because they could potentially generate income in the future; however, more time is required to realize this opportunity. A study focusing on developing social entrepreneurship and, hopefully, matching mentors could be useful for refugees. It can also be helpful to focus a study on the challenges refugee women perceive before entrepreneurship.

The study did not involve any representatives from the government; therefore, our interpretation of the policies at the societal level is limited by the studies of other researchers (secondary data). Our study did not consider the legitimacy of our participants' claims to stay in Hong Kong. The five interviewees were asylum seekers; the status of the other four participants may or may not be refugees. Therefore, we cannot stratify the findings to a single group: asylum seekers. From our perspective, education is a human right. If they are in Hong Kong, they should be given a platform to benefit from education and, in turn, benefit the city following this type of education. However, this could be a future discussion and research area. We suggest future research to consider whether and how RAS' reasons for seeking asylum or status influence their investment in education.

Our study suggests that generic green skills and education for sustainable development pedagogy work well together and that social enterprises could become a platform to utilize the skills learned. We recommend conducting similar research with different groups who have work permits in Hong Kong and who lack empowerment, such as minority youth and/or victims of domestic abuse. Research with these groups could add value to academic literature and have practical implications for people's lives and Hong Kong's sustainable urban development.

References

Aerne, A., & Bonoli, G. (2021). Integration through vocational training. Promoting refugees' access to apprenticeships in a collective skill formation system. *Journal of Vocational Education & Training*. https://doi.org/10.1080/13636820.2021.1894219.

Aksoy, C. G., Ginn, T., & Malpassi, F. (2024, May 15). *Attitudes and policies toward refugees: Evidence from low- and middle-income countries*. Retrieved from https://www.jointdatacenter.org/wp-content/uploads/2024/09/Ginn-1.pdf

Al-Dajani, H. (2022). Refugee women's entrepreneurship: Where from and where to next? *International Journal of Gender and Entrepreneurship, 14*(4), 489–498. https://doi.org/10.1108/IJGE-06-2022-0090

Alsop, R., Bertelsen, M. F., & Holland, J. (2006). *Empowerment in practice from analysis to implementation*. The World Bank. Retrieved from https://elibrary.worldbank.org/doi/pdf/10.1596/978-0-8213-6450-5

Arbuthnott, K. D. (2009). Education for sustainanle development beyond attitude change. *International Journal of Sustainability in Higher Education, 10*(2), 152–163.

Beaman, L., Onder, H., & Onder, S. (2022, March). When do refugees return home? Evidence from Syrian displacement in Mashreq. *Journal of Development Economics, 155*. https://doi.org/10.1016/j.jdeveco.2021.102802.

Biesta, G. (2009). Good education in an age of measurement: On the need to reconnect with the question of purpose in education. *Educational Assessment, Evaluation and Accountability, 21*, 33–46. https://doi.org/10.1007/s11092-008-9064-9

Bratman, G. N., Hamilton, J. P., Hahn, K. S., & Daily, G. C. (2015). Nature experience reduces rumination and subgenual prefrontal cortex activation. *Psychological and Cognitive Sciences, 112*(28), 8567–8572. Retrieved from https://www.pnas.org/doi/10.1073/pnas.1510459112

Chadderton, C., & Edmonds, C. (2014). Refugees and access to vocational education and training across Europe: A case of protection of white privilege? *Journal of Vocational Education & Training, 67*(2), 136–152. Retrieved from https://www.tandfonline.com/doi/full/10.1080/13636820.2014.922114?src=recsys

Constable, N. (2014). *Born out of place: Migrant mothers and the politics of international labor*. University of California Press. Retrieved from https://web.b.ebscohost.com/ehost/ebookviewer/ebook/bmxlYmtfXzcwODE4N19fQU41?sid=0f0b20c1-3d35-42ecbb92-bcdb578ef02f@pdc-v-sessmgr03&vid=0&format=EB&rid=1

Freire, P. (1970). *Pedagogy of the oppressed*. (M. B. Ramos, Trans.). Penguin Books.

HKSARG. (2019). *Report on manpower projection to 2027*. Labour and Welfare Bureau. Retrieved from https://www.lwb.gov.hk/en/other_info/mp2027_en.pdf

Illinois Coalition for Immigrant and Refugee Rights. (2006, December). *For the benefit of all*. Retrieved from https://core.ac.uk/download/71339306.pdf.

ILO Office for Turkey. (n.d.). *Transition to formality programme: Phase one impact analysis*. ILO Office for Turkey, SGK and T.C Aile, Calisma ve Sosyal Hizmetler Bakanligi. Retrieved from https://www.ilo.org/wcmsp5/groups/public/%2D%2D-europe/%2D%2D-ro-geneva/%2D%2D-iloankara/documents/publication/wcms_763393.pdf

Kennedy, K. J. (2011). The "long march" toward multiculturalism in Hong Kong: Supporting ethnic minority students in a Confucian state. In J. Phillion, M. T. Hue, & Y. Wang (Eds.), *Minority students in East Asia: Government policies, school practices, and teacher responses* (pp. 155–173). Routledge.

Lee, G. K., & Chan, E. H. (2009). Indicators for evaluating environmental performance of the Hong Kong urban renewal projects. *Facilities, 27*(13/14), 515–530.

Loper, K. (2010). Human rights, non-Refoulement and the protection of refugees in Hong Kong. *International Journal of Refugee Law, 22*(3), 404–439. Retrieved from https://hub.hku.hk/bitstream/10722/92989/2/Content.pdf

Mathews, G. (2018). Asylum seekers as symbols of Hong Kong's non-Chineseness. *China Perspectives, 18*(3), 51–58. https://doi.org/10.4000/chinaperspectives.8132

McGranahan, G., Marcotullio, P., Bai, X., Balk, D., Braga, T., Douglas, I., et al. (2005). Urban systems. In *Ecosystems and human Well-being, current state and trends* (Vol. 1, pp. 795–825). IIED. Retrieved from https://www.iied.org/g00505

Narayan, D., & Petesch, P. (2007). Agency, opportunity structure, and poverty escapes. In D. Narayan & P. Petesch (Eds.), *Moving out of poverty: Cross-disciplinary perspectives on mobility* (Vol. 1, pp. 1–44). Palgrave Macmillan and the World Bank. https://doi.org/10.1596/978-0-8213-6991-3

New American Economy. (2017). *From struggle to resilience: The economic impact of refugees in America*. New American Economy. Retrieved from https://www.newamericaneconomy.org/wpcontent/uploads/2017/06/NAE_Refugees_V5.pdf

Pavlova, M. (2012). Towards using transformative education as a benchmark for clarifying differences and similarities between environmental education and education for sustainable development. *Environmental Education Research, 19*(5), 656–672. https://doi.org/10.1080/13504622.2012.736476

Pavlova, M. (2022). Introduction: Unpacking greening and skills recognition in micro, small, and medium enterprises. In I. M. Pavlova & M. Singh (Eds.), *Recognizing green skills through non-formal learning. Education for sustainability* (Vol. 5, pp. 3–22). Springer. https://doi.org/10.1007/978-981-19-2072-1_1

Pavlova, M., & Chen, C. S. (2019). Facilitating the development of students' generic green skills in TVET: An ESD pedagogical model. *The Online Journal for Technical and Vocational Education and Training in Asia, 12*, 1–20. Retrieved from http://tvet-online.asia/issue/12/pavlova-etal/

Peri, G. (2012). The effect of immigration on productivity: Evidence from US states. *The Review of Economics and Statistics, 94*(1), 348–358. Retrieved from https://www.jstor.org/stable/41349180?seq=9#metadata_info_tab_contents

Sen, A. (1999). *Development as freedom*. Oxford University Press.

Startup Refugees. (2022). *What is startup refugees*. Retrieved from Startup Refugees. https://startuprefugees.com/about/

Task Force. (2020). *VPET 3A for the future: Task force promotion of vocational and professional education and training*. The Education Bureau.

Taylora, J. E., Filipskib, M. J., Alloush, M., Gupta, A., Rojas Valdes, R. I., & Gonzalez-Estradac, E. (2016). *Economic impact of refugees. Economic Sciences, 113*(27), 7449–7453. https://doi.org/10.1073/pnas.1604566113

The Republic of Poland. (2022, March 23). *Important information for the citizens of Ukraine interested in working in Poland*. Retrieved January 25, 2025, from News. https://www.gov.pl/web/family/important-information-for-thecitizens-of-ukraine-interested-in-working-in-poland

Tiwari, S. (2021). Do macroeconomic fluctuations at destination matter in determining migrants' return decisions? *World Development, 148*. https://doi.org/10.1016/j.worlddev.2021.105662

UN-Habitat. (2017). *New urban agenda. Habitat III secretariat.* UN-Habitat. Retrieved from http://habitat3.org/wp-content/uploads/NUA-English-With-Index-1.pdf

UNHCR. (2019). *Global framework for refugee education. UNHCR.* Retrieved from https://www.unhcr.org/events/conferences/5dd50ce47/global-refugee-forum-pledging-guidance-global-framework-refugee-education.html

UNHCR. (2025a). *Durable Solutions.* Retrieved from UNHCR: https://www.unhcr.org/sites/default/files/legacy-pdf/61a0fe634.pdf

UNHCR. (2025b). *Technical and vocational education and training.* Retrieved from What we do: https://www.unhcr.org/what-we-do/build-better-futures/education/higher-education-and-skills/technical-vocational-education-training

UNHCR Poland & Deloitte Polska. (2024, March 5). *Study finds positive impact of refugees on Poland's economy.* Retrieved from News. https://www.unhcr.org/europe/news/news/study-finds-positive-impact-refugeespoland-s-economy

Vecchio, F. (2016). *Asylum seeking and the Global City.* Routledge.

Verkuyten, M. (2021., January 12). Public attitudes towards migrants: Understanding cross-national and individual differences. *World Pscyhiatry, 20*(1), 132–133. https://doi.org/10.1002/wps.20819

Verme, P. (2023, March 28). *Theory and evidence on the impact of refugees on host communities.* Retrieved from World Bank Blogs. https://blogs.worldbank.org/en/dev4peace/theory-and-evidence-impact-refugees-hostcommunities

Wanner, T. (2015). The new 'passive revolution' of the green economy and growth discourse: Maintaining the 'sustainable development' of neoliberal capitalism. *New Political Economy, 20*(1), 21–41. https://doi.org/10.1080/13563467.2013.866081

Appendices

Appendix A. Description of the participants

Interviewee 1: Ms. Iris[1]

Originally from an African country, Ms. Iris has two sons. She came to Hong Kong as a visitor but got stuck there seven years ago at the time of the interview and has been waiting for a decision about her status since then. She lives as a single parent with one of her sons in a tiny living space. The other son is with the social welfare department. The child's father is from another African country and is also an asylum seeker in Hong Kong. He was reported to have been abusive.

Ms. Iris said she has international work and travel experience. Before coming to Hong Kong, she worked in her country of origin, Singapore, Oman, China, and Dubai. For example, she worked as a clerk in her home country in a construction company. She also worked at a supermarket in Dubai. Her main problem is housing in Hong Kong. She believes the regulations concerning employment put refugees in a jeopardous position because if they are not involved in illegal activities, such as prostitution and/or drug dealing, they cannot survive. Faith/religion helps her cope with her daily life.

Our workshops were the first-ever educational event she had participated in during her seven years in Hong Kong.

Interviewee 2: Ms. Lilly

Ms. Lilly is from a South Asian country and is a single parent with two children.

She attended college but dropped out after the first year for economic reasons. She was majoring in business/trade. She has work experience in the motherland and abroad (in Singapore) as part of a teaching crew, a salesperson in the cosmetic sector, and in a multinational fast-food restaurant chain.

[1] Pseudonyms are used.

A. T. Saral, M. Pavlova, *Education for Refugees and Sustainable Green Urban Development*, Education in the Asia-Pacific Region: Issues, Concerns and Prospects 78, https://doi.org/10.1007/978-981-95-1680-3

She came to Hong Kong as a domestic helper in 2007. Then, she became pregnant out of wedlock with her partner, who is also an asylum seeker. She applied for asylum in 2014 and is still waiting for a result. Her cousin also lives in Hong Kong with her family. At the time of the interview, she was staying in a shelter house to escape physical abuse from her partner. Although she lacks status, financial aid, and housing issues challenge her life, she prefers to stay in Hong Kong for her children's future.

Interviewee 3: Ms. Rose

A South Asian-origin single mother, Ms. Rose lives with her daughter in Hong Kong. She has two sons back in her country of origin. Her family was big, back home, and involved in farming. She grew up on the farm and learned about plants and animals. She began studying business administration at university in her home country but dropped out in her second year. She came to Hong Kong in 2002 as a domestic helper and applied for asylum because her daughter's father is an asylum seeker in Hong Kong. She wants to remain there for her daughter's schooling.

Lack of employment and housing-related issues are her main problems in Hong Kong. She wants to work and earn a living rather than depending on the government and NGOs.

Interviewee 4: Ms. Peony

Ms. Peony is an asylum seeker from South Asia. She is a single mother and a victim of abuse. She is generally satisfied with living in Hong Kong. Help from the government is insufficient but appreciated, and not all government-provided items are of good quality. Therefore, she valued the coupons from our workshop. In addition to legal issues, such as going to the Immigration Department, to court, etc., she is attempting to manage her and her daughter's life on the small subsidy, and she reported this was challenging. She would prefer to be able to work rather than be dependent on government assistance. However, she also fears being incarcerated (although she was not working at the time of the interview).

Ms. Peony came to Hong Kong as a domestic helper. Before that, she worked at a company's cash registry desk. She studied at university but dropped out after the second year. Although she was offered a sponsorship to continue her education in Hong Kong, she could not do it due to a lack of time and daily life constraints.

She prefers to stay in Hong Kong for her daughter's education.

Interviewee 5: Ms. Daisy

Ms. Daisy is a single mother of South Asian origin. She comes from a very large family, but only two of her 10+ siblings live. They all died for various reasons, but they were mainly health related. She also has health issues, such as dizziness.

She came to Hong Kong as a domestic helper in the early 2000s. Before that, she also worked as a domestic helper in Saudi Arabia. She is 50+ years old and gave birth to her son when she was 43. She had already applied for asylum in Hong Kong before expecting the baby. The child's father lives abroad and sends money monthly so their son can attend tutorials in Chinese and other lessons. As a result, Ms. Daisy's financial situation is better than that of other interviewees.

The primary stressor for her is being in limbo due to her legal status. She worries that she cannot take her son with her if she returns home. However, even if the court decides against her claim, she is accustomed to appealing the decision. She is not keen on working due to her age and ill health.

Most refugees and asylum seekers we interviewed have been in Hong Kong for many years and have prior working experience abroad or in Hong Kong as domestic helpers. They did not complete their university degrees for economic reasons. They have children and are victims of abuse from their partners. Their legal status, housing, and employment issues stress them daily. However, they are willing to stay in Hong Kong despite potential difficulties because they value their children's future and education.

Interviewee 6: Mr. Indigo—Founder of a Social Enterprise

Mr. Indigo is originally from an African country and came to Hong Kong in 2014 as an asylum seeker. After he arrived, he studied at a university. He is the first refugee of African origin to have graduated from a university in the city.

He shared that, as a refugee, he experienced hard times in the beginning and was tricked by people who said they could help him. He commented that the system in Hong Kong is better now in supporting RAS than when he arrived. He thinks society's expectations about refugees are very low. However, RAS needs to realize how resilient they are and demonstrate this to the government and people of Hong Kong instead of expecting help from society or crying for help.

He started four startups in the city; two failed, but the others are running with other people. He works at a bank and founded a well-known social enterprise in Hong Kong. The center's mission is to rebrand blackness and change the perception of Africans in Asian eyes. Hong Kong is the starting point for this mission. Book clubs and workshops, such as drumming and cooking workshops, arranging cultural tours in the city, and brown bag seminars, are among the many activities they organize. When discussing his success, he mentioned the importance and difficulty of receiving an honest opinion about a product or service they produce. People may not be honest because they feel pity. He shared his vision about founding businesses—instead of being concerned only about himself and his current problems, he focuses on the future and the children's children.

Mr. Indigo is also invited as a consultant to give talks to various schools that would like to adopt a multicultural approach and on platforms such as TEDx.

Interviewee 7: Mr. Aaron—Representative from the Branches of Hope (BoH).

Background of the organization: The mission of Branches of Hope is to help refugees with their daily needs while at the same time facilitating activities for local people to engage with this group. In this way, they contribute to integrating the refugee community within Hong Kong. They have two initiatives, STOP and ROAD; STOP, which is about human trafficking; and ROAD, which is about refugee opportunities and development. At the time of the interview, the NGO's representative, Mr. Aaron, has been a case officer at BOH for over a year. He is involved in the ROAD program, where he communicates with different parties about refugees' daily requirements, such as legal issues, housing, and medical needs.

The organization runs several educational programs; some are long-term, and some are short-term. For instance, an English language course is long-term for adults, and BoH facilitates one-to-one tutorial lessons. Volunteers or teachers at international schools provide these lessons. In addition, they hold extra-curricular classes such as singing and dancing for children, as well as computer lessons for adults. In the past, they held a mechanics class for a short period, which was facilitated by a refugee who had been a mechanical engineer back in his homeland. However, due to COVID-19, turnover rates decreased. As a result, these educational programs were put on hold for a while. As a case officer in the ROAD program, Mr. Aaron mentioned that environmental issues are not the primary concern in their mind.

For this reason, our initiative was new to the organization. BoH was motivated to collaborate due to the refugees' positive reaction to our Green Refugee Day.

Mr. Aaron commented that refugees are stressed daily, and their well-being is influenced by multiple factors, such as housing, lack of employment, complicated legal issues concerning their claims, and bad depictions in media and the eyes of society. However, he mentioned improvements in the system, particularly in employment. While NGOs are there to support this community, RAS understand and support one another and are willing to participate in activities because they like to learn skills and try new things.

Appendix B Survey Questions

Section 1: General Opinion

The questions in this section are about your general opinion. Please rank your answers for the items below.

1. What are the **four most important issues** you are personally facing at the moment?

   (1—Most important, 4—Least important)

   You may rank up to four.

   - Housing
   - Unemployment
   - Rising prices/Cost of living
   - Environmental problems (e.g. air quality, wastes etc.)
   - Health provision
   - Education system
   - Economic situation
   - Assistance by the ISS Bureau HK
   - Duration of claim check/status approval
   - Taxation
   - Crime
   - Political instability (e.g. protests)
   - None
   - Don't know
   - Other (Please specify: …)

2. Overall, how satisfied are you with your life now?

   - Very satisfied
   - Fairly satisfied
   - Not very satisfied
   - Not at all satisfied
   - Don't know*Please indicate to what extend you agree or disagree with following statement:*

3. I am a part of the Hong Kong society.

   - Strongly agree
   - Agree
   - Neutral
   - Disagree
   - Strongly disagree

Section 2: General attitudes and behavior concerning environment

The questions in this section are about your attitudes and behavior concerning environment.

4. Do you think protecting the environment is important?

   - Very important
   - Fairly important
   - Not very important
   - Not at all important
   - Don't know

Please rank your choices for the items below.

5. What are the eight most important **environmental problems** facing Hong Kong at the moment?

   (1 – Most important, 8- Least important)
   You may rank up to eight.

   - Climate change
   - Air pollution
   - Marine pollution
   - Pollution of rivers, lakes and ground waters
   - Growing amount of waste
   - Waste management (e.g. Recycling)
   - Decline or extinction of species, habitats and ecosystems (e.g. forests, fertile soils)
   - Agricultural pollution (e.g. use of pesticides, fertilizers etc.) and land management (e.g. desertification)
   - Shortage of drinking water
   - Frequent droughts or floods
   - Noise pollution

- Over consumption
- None
- Don't know
- Other (Please specify: …)

6. From the following list, please **rank three main sources of information for you** about the environment.
   (1 – Most used source of information, 3 – Least used source of information)
   You may rank up to three.

- Television news
- The internet (e.g. websites, blogs, forums, etc.)
- The radio
- Films and documentaries on television
- Online social networks
- Newspapers
- Family, friends, neighbors or colleagues
- Other (Please specify: …)

7. In your opinion, which of the following would be the most effective ways of tackling environmental problems in Hong Kong? Please rank your answers up to six choices.
   (1 – Most effective, 6-Least effective)
   You may rank up to six.

- Changing the way we consume
- Changing the way we produce and trade
- Investing in research and development to find technological solutions
- Providing more information and education, e.g. on waste separation, energy consumption
- Introducing stricter environmental legislation
- Introducing heavier fines for breaches of environmental legislation
- Encouraging businesses to engage in sustainable activities
- Making the food system more sustainable from production to consumption
- Using financial incentives (e.g. tax breaks, subsidies) to protect the environment
- Ensuring better enforcement of legislation
- Making the banking and insurance systems more environmentally friendly.
- None
- Don't know
- Other (Please specify: …)

8. In your opinion, who is responsible for solving environmental problems?
   (1 – Most responsible, 5 – Least responsible)
   You may rank up to five.

- Government
- Private Companies/Enterprises

- Non-governmental organizations (i.e. NGOs)
- All households
- My household
- Combination of several (Please specify: …)
- Other (Please specify: …)

Please indicate to what extend you agree or disagree with following statements:

9. Environmental issues have a direct effect on my daily life and health.
   - Strongly agree
   - Agree
   - Neutral
   - Disagree
   - Strongly disagree
10. Hong Kong as a whole is generating too much waste.
   - Strongly agree
   - Agree
   - Neutral
   - Disagree
   - Strongly disagree
11. My household is generating too much waste.
   - Strongly agree
   - Agree
   - Neutral
   - Disagree
   - Strongly disagree
12. I make efforts to reduce my waste.
   - Strongly agree
   - Agree
   - Neutral
   - Disagree
   - Strongly disagree

- *If you selected "neutral, disagree, strongly disagree", please go to question 12.1. If you selected "strongly agree or agree", please skip to question 13.*

  12.1 If you selected "neutral, disagree, strongly disagree" in previous question, please select the reasons **why you are not trying** to reduce wastes. *You may choose multiple answers.*

- Reducing waste is not important to me.
- I don't know how I can reduce waste.
- It is the responsibility of the product producer to reduce waste, not mine.
- It is the responsibility of the government to stimulate my waste reduction.

- The level of waste I generate is already at its minimum, I can't reduce more.
- I tend to throw things away as it is difficult or too expensive to get them repaired.
- Don't know
- Other (Please specify: …)

13. How would you judge the current waste management system in Hong Kong?
    - Very efficient
    - Efficient
    - Inefficient
    - Very inefficient
    - Don't know

14. What type of bag do you usually use when shopping?
    - Plastic bag
    - Paper bag
    - Textile bag
    - Other (Please specify: …)

- *If you selected "plastic bag", please go to question 14.1. If you selected "textile bag", please skip to question 14.2.*

14.1 If you use plastic bag, what do you do with it afterwards?

- Throw it away
- Re-use it
- Other

- (Please specify: …)

14.2 If you use a textile bag, what do you do to remember bringing it with you?

- Store it in your shoulder bag
- Hang it by the door
- Keep them in your clothes' pocket
- Other

- (Please specify: …)

15. Do you have a collection point for recyclable items near where you live?
    - Yes
    - No
    - Don't know

16. Please mark the other areas where you have noticed recycling points.
    You may choose multiple answers.
    - School/Kindergarten
    - Community center/shopping mall
    - Park

- Hospital
- Other (Please specify: …)

Below are the images of everyday products.

17. Please mark the items you recycle below.
    You may choose multiple answers.

- Cans
- Plastic bottles
- Glass bottles
- Plastic cups
- Plastic food containers
- Plastic bags
- Newspapers
- Paper bags
- Carton boxes
- Aluminum food containers
- None
- Other (Please specify: …)

18. From A to F, how would you judge your consumption habits?
    A—I am very conscious about the environmental impacts of what I eat, buy or use.
    F—I do not consider the environmental impact of what I eat, buy or use.)
    You may circle the most fitting number for your answer.

Please indicate to what extend you agree or disagree with following statements:

19. My consumption habits effect the environment in Hong Kong and the rest of the world.

    - Strongly agree
    - Agree
    - Neutral
    - Disagree
    - Strongly disagree

20. I would like to learn how I could protect the environment with my daily activities.

    - Strongly agree
    - Agree
    - Neutral
    - Disagree
    - Strongly disagree

21. I would like to take part in activities organized by different organizations to provide environmental solutions for Hong Kong.

    - Strongly agree
    - Agree
    - Neutral
    - Disagree
    - Strongly disagree

22. Please indicate how often you do activities below.

	Always	3+ times a week	1–3 times a week	1–2 times a month	1–2 times in six months	Never
I separate plastics, papers and glasses for recycling.	O	O	O	O	O	O
I re-use plastics products to avoid single-use (e.g. re-use plastic bags, refill my own water bottle)	O	O	O	O	O	O
I cut down my energy consumption by turning down air-conditioning, not leaving appliances on stand-by or buying energy-efficient products.	O	O	O	O	O	O
I speak to others about environmental issues.	O	O	O	O	O	O
I repair a product instead of replacing it.	O	O	O	O	O	O
I avoid buying over-packaged goods.	O	O	O	O	O	O

I buy or use second hand products instead of new ones such as clothes or electronics.	O	O	O	O	O	O
I join environmental demonstrations, attend environment-related workshops, take part in environmental activities (e.g. a collective beach or park cleanup)	O	O	O	O	O	O
I cut down my water consumption such as taking showers in less than 5 min or using less water to wash dishes.	O	O	O	O	O	O
I use rechargeable batteries.	O	O	O	O	O	O
I make home composting.	O	O	O	O	O	O
I buy exactly what I need to avoid food waste and other types of wastes.	O	O	O	O	O	O
I avoid buying products that pollute air such as hair spray and air freshener.	O	O	O	O	O	O
I warn someone who litters.	O	O	O	O	O	O
I talk to my kids about protecting the environment.	O	O	O	O	O	O
I share links on social media concerning environment.	O	O	O	O	O	O
I follow an environmental organization online (such as UNEP, WWF etc.).	O	O	O	O	O	O
I donate/sell the items I no longer need such as clothes, plates, books etc.	O	O	O	O	O	O
I dispose household batteries and light bulbs in designated areas.	O	O	O	O	O	O
Other (please specify: …)	O	O	O	O	O	O
Other (please specify: …)	O	O	O	O	O	O
Other (please specify: …)	O	O	O	O	O	O

Section 3: Environmental values

The questions in this section are to understand your environmental values. *Please complete the sentences below with one of the options provided.*

23. Protecting the environment is important …… **(Please complete with one of the options below.)**
    - It contributes to pleasure and wellbeing of humans.
    - It is home to many plant and animal species.
    - Other (please specify:…)

24. We need to keep lakes and rivers clean in order to …… **(Please complete with one of the options below.)**
    - Have a place to enjoy water sports and do leisure activities.

- Maintain marine ecology and biodiversity.
- Other (Please specify:…)

25. We need to protect endangered species because …… **(Please complete with one of the options below.)**

- They may be useful for humans in the future.
- They have a right to live.
- Other (Please specify:…)

26. Natural resources are important because …… **(Please complete with one of the options below.)**

- They provide food, water and shelter for living organisms such as animals and plants.
- They provide commercial and industrial products for economic growth.
- Other (Please specify:…)

27. Environmental problems should be addressed to ensure …… **(Please complete with one of the options below.)**

- Countries' economic development.
- Biologically diverse and productive ecosystems.
- Other (Please specify:…)

Section 4: Social enterprises*

The questions in this section are related to social enterprises.

**Social enterprise is a type of business that tries to address societal and environmental problems while generating income.*

28. Have you ever started your own business or participated in an enterprise?

- Yes
- No

29. In terms of environmental sustainability and social prosperity, which of the following actions, *if any*, should an enterprise you wish to work at or establish one day actively take? *Please rank six most important actions.*

(1- **Most important, 6—Least important)**

You may rank up to six.

- Recycling or reusing materials
- Reducing consumption of natural resources (e.g. saving water or switching to sustainable resources)
- Saving energy or switching to sustainable energy sources
- Improving working conditions of its employees
- Promoting and improving diversity and equality in the workplace
- Evaluating the impact of the enterprise on society
- Engaging employees in the governance of the enterprise
- Developing environmentally friendly products or services

- Buying environmentally friendly products or services for the business
- Monitoring and managing the amount of waste, energy and water generated
- Having a risk management system
- Having an impact assessment on the environment
- None
- Don't know
- Other (Please specify: ...)

30. If you could set up a social enterprise in Hong Kong, which of the following areas would pose the biggest challenges for you?

You may choose maximum 3 answers.

- Regulatory obstacles or administrative burden
- Lack of financial resources
- Lack of awareness how to set it up
- Skills, including managerial skills
- Difficulties with digitalization and innovation
- Access to information
- Lack of support from friends/family
- Lack of consumer or customer demand
- Lack of interest to set up a social enterprise
- No challenge
- Don't know
- Other (Please specify: ...)*Please indicate to what extent you agree or disagree with the following statements:*

31. I believe social enterprises can help with finding a solution to environmental problems.
    - Strongly agree
    - Agree
    - Neutral
    - Disagree
    - Strongly disagree
32. I would like to support environmental efforts of Hong Kong society.
    - Strongly agree
    - Agree
    - Neutral
    - Disagree
    - Strongly disagree
33. If allowed, what would be your motivation to set up a social enterprise?
    - To generate income
    - To contribute to environmental solutions
    - Both
    - Other (Please specify: ...)

Section 5: Demographics

The questions in this section are about your background.

34. Please specify your country of origin.

……………………………………………………………………………

35. What is your gender

- Female
- Male

36. What is your age group?

- 18–24
- 25–34
- 35–44
- 45+
- Prefer not to say

37. What is the highest level of education you have completed?

- Some high school
- High school
- Vocational/trade school
- Bachelor's degree
- Graduate degree (e.g. Master's, PhD)
- Other (Please specify: ...)
- Prefer not to say

37. Are you married?

- Yes
- No
- Prefer not to say

38. How many children do you have?

- 0
- 1
- 2
- 3
- 4 or 4+
- Prefer not to say

39. How would you describe your economic status?

- I usually have difficulties paying the bills.
- I sometimes have difficulties paying the bills.
- I rarely/almost never have difficulties paying the bills.

40. Please select the workshops you participated in.
    - Sustainable consumption—toy & Wallet Making from leftover fabric
    - Waste management—Soap Recycling & Candle Making from used cooking oil
    - Social Enterprise I—waste glass bottle painting
    - Social Enterprise II—Face & Body Scrub from used coffee grounds
41. What part(s) of the workshop was the most effective in developing your skills and understanding?
    - Information about sustainability concepts
    - Case study examples
    - Activities about the concepts and case studies
    - Information about social enterprises
    - Hands-on activities (toy making, soap recycling etc.)
    - Involving kids in the activity
    - Doing an activity with other mothers
    - Other (Please specify: …)
42. Please write down which activity you liked the most and why?
43. Which skills you learned in the workshops can you apply in daily life?
44. How can we improve our workshops? Please write your recommendations below.

Appendix C. Interview Guidelines for Workshop Participants, NGO Representative, and SE-Founder

- **Interview Guideline for Workshop Participants**

Section 1: General Opinion

1. Overall, how do you feel about your wellbeing in Hong Kong?
   (a) Are you satisfied with your living here?
2. What do you think are the problems influencing your wellbeing in Hong Kong at the moment?
   (a) How different are these problems from your experience in your country of origin?
3. What are your needs and expectations for a better life?
   (a) How can Hong Kong society and government can support you?

Section 2: General attitudes and behavior towards environment

1. Do you think environmental problems influence your life and wellbeing?
   (a) If so, in what ways?

2. What are the environmental problems that draw your attention in Hong Kong?
    (a) Are they different than in your hometown?
        (i) Have you paid attention to environmental issues back home?
3. Do you think one person's actions influence the environment? (For example, use of electricity, water, disposal of used items etc.)
    (a) How do you think your actions influence the environment?
4. Are you aware of policies and practices in Hong Kong about the protection of environment?
    (a) What are the differences in terms of environmental practices and policies between HK and your home country?
5. Are you concerned with reduction of waste, recycling of materials and reuse of items in your daily life?
    (a) Can you give an example of reuse, reduce, and recycling actions you perform?
        (i) How often do you do these?
6. Do you try to teach your kids about protecting the environment and nature?
    (a) What do you usually tell/teach your kids about protecting the environment and nature?
7. Do you know if your kids learn about environmental issues at school?
    (a) Do you think it is important? What is your reaction to it?
8. In what ways do you think environmental problems are influencing or will influence them?

Section 3: Environmental values

1. Do you think we should protect the environment?
    (a) Why? Why Not?

Section 4: Social Enterprises

1. Do you have any experience working in Hong Kong or before coming to Hong Kong?
    (a) What was the nature of your job and place?
2. Are you aware of Hong Kong's policy towards social entrepreneurship?
3. What does social entrepreneurship mean to you?
    (a) Do you think we can achieve social change via Social Enterprises?
        (i) How would this change benefit the environment?

4.Do you see yourself contributing to your community or Hong Kong society in general?

   (a) In what ways?

5. Would you consider working in a social enterprise or setting up your social enterprise?

   (a) How would this influence your wellbeing?
   (b) What would be your motivation to set up a social enterprise?

6. What do you think would be the challenges before setting up a social enterprise for you?

   (a) How can Hong Kong government and society support you?

7. What do you expect to gain from workshops?

   (a) Do you have any suggestions for such activities? What do you think could be a good recycle/reuse/repurpose activity?

- **Interview Guideline for NGO representative**

1. Could you please tell us a bit about your organization?

   (a) What type of activities do you organize?

      (i) What are the social and environmental impacts of your organizations' activities?

   (b) Who is your target group?

      (i) How would you define the wellbeing of this group?
      (ii) What are the needs of the concerned group in HK?
      (iii) What do you think are the main challenges in meeting these needs?

2. What was your organization's motivation to participate in our study?
3. In what ways do you think our study benefits the concerned community, your organization and HK society?
4. SE is one of the concern areas of this study. What are the social and environmental benefits of SEs from your perspective?
5. Can disadvantaged groups such as ethnic minorities, refugees, and asylum seekers set up a SE in HK?

   (a) What are the regulations that ease or hamper the process for different groups?
   (b) Are there any SEs that address the needs of disadvantaged groups in HK?

      (i) What is the nature of their activity? Do they concentrate on low-level or high skilled work?

6. Considering the needs in market and skills gap in HK, in what areas do you think the disadvantaged groups can contribute to Hong Kong?

7. What would be your advice to those who would like to set up a SE with environmental benefits?
   (a) Are there any (international or local) grants that refugees and asylum seekers can apply to?
8. For someone who has been involved in charitable and social work for a long time, have you observed positive changes for people working or volunteering in an organization or a SE?
   (a) In what ways do you think participation in social and environmental events can benefit disadvantaged groups' wellbeing?
9. What can be done on a governmental or societal level to support disadvantaged groups?

- **Interview Guideline for SE-founder**

1. Can you tell us briefly about your background?
   (a) What was your experience like when you arrived Hong Kong as a refugee?
2. What was your motivation to set up a social enterprise?
3. Can you tell us step-by-step how you established your enterprise?
4. Did you have any challenges before setting up a social enterprise as a refugee?
   (a) How did you overcome these challenges?
5. What would be your advice to those who would like to set up an enterprise in Hong Kong?
6. Now that your enterprise has been functioning for over a year, what benefits has your social enterprise brought to individuals?
   (a) How about the benefits for the community that you are concerned with? (*Black community)
   (b) In addition, the benefits to HK society in general?
7. How do you promote your services?
   (a) Do you use contacts at individual, governmental or institution level?
8. Based on your experience, how important do you think partnership is for a successfully run social enterprise?

What do you think HK government or society should do to support refugee groups?

Appendix D Observation form for the Workshops (To-Be-Used by the Student Helper and the Researchers)

<table>
<tr><td colspan="2">1) What do you think about refugees' motivation to participate in the workshop?
1.1 Did it increase or decrease at certain stages of the workshop?
2) Do you think refugees had the opportunity to develop the following skills?</td></tr>
<tr><td>Week 2 Sustainable consumption</td><td>Cognitive competencies:
• Environmental awareness and a willingness to learn about sustainable development
Technological competences:
• Material use and impact quantification</td></tr>
<tr><td>Week 3
Waste management</td><td>Cognitive competencies:
• Systems and risk analysis, skills to assess, interpret and understand both the need for change and the measures required for greening at different levels (namely state, enterprise and personal)
• Understanding the complexity and interconnectedness of sustainable development issues and challenges
Interpersonal skills:
• Communication and negotiation skills to discuss conflicting interests in complex and greening contexts
Intrapersonal competencies:
• Adaptability and transferable skills that help mothers learn and apply the new technologies and processes required to green their homes (-and green their jobs, if allowed to work, in the future)</td></tr>
<tr><td>Week 4 / 5
Social Enterprise I/II</td><td>Cognitive competencies
• Ability to think about things differently
• Identifying ways of being part of the solution
Interpersonal competencies
• Coordination, management and business skills to facilitate holistic and interdisciplinary approaches that encompass economic, social and ecological objectives
• Marketing skills to promote products and services
Intrapersonal competencies
• Adaptability and transferable skills that help mothers learn and apply the new technologies and processes required to green their homes (-and green their jobs, if allowed to work, in the future)</td></tr>
<tr><td colspan="2">3) Have you noticed any other green skills that refugees possess –based on the above explanation?
3.1 What do you think is their level? (high, medium, low)
4) How would you describe the interaction between Aysuhan and participants; participants and the kids; among participants?
5) In what ways do you think the workshops can be improved for the next time?</td></tr>
</table>

Appendix E Observation form for the Group Discussion (To-Be-Filled by NGO Representatives)

Study Title
Date
Start time
End time
Moderator name
Observer's name
Group detail
No. participants
Participant demographics
Gender/age/ job
Reflection of the observer on the following:
1. General opinion *Please write down your initial reflection on participants' motivation to join the workshops.* • *What were their motivations to come?* • *What is their opinion about the workshop?*
2. General attitude and behavior towards environment *Please write down your initial reflection concerning the group's understanding of their contribution to environment, change in attitude and/or behavior as a result of workshops.* • *Did the activities from workshops seem doable in daily life from their perspective? (Please refer to some examples they gave)* • *Did participating in the workshop indicate any change in their knowledge, behavior and attitude towards the environment? (Please refer to some examples they gave)*
3. Environmental values *Please write down your initial reflection about participants indicated environmental values and environmental impact on wellbeing.* • *Where do participants see the environmental problems effect in their life?* • *What is their preferred position to care about the environment: For its own sake, or for human's sake?*
4. Social enterprises *Please write down your initial reflection on participants' motivation to set up a social enterprise if allowed.* • *Did participants seem motivated about setting up a social enterprise (if allowed)?* • *What seemed to be their motivation? Environment/social/economic etc.?* • *Would they like to learn more about environmental issues?*
5. About focus group *Please write down your initial reflection about the conduct of focus group discussion.* • *What was the level of interest in terms of engagement in the discussion?* • *What topics were particularly engaging for participants?* • *How was the rapport between Aysuhan and participants?* • *Was it an open interaction?* • *What does the participants' body language tell you?* • *Can you see any indication if they would like to be involved in such activities in the future?*
6. Additional thoughts and ideas *Please write down your additional thoughts and ideas concerning the focus group discussion below.*

Appendix F. Social Business Canvas of Participants

Team 1—Social business canvas.

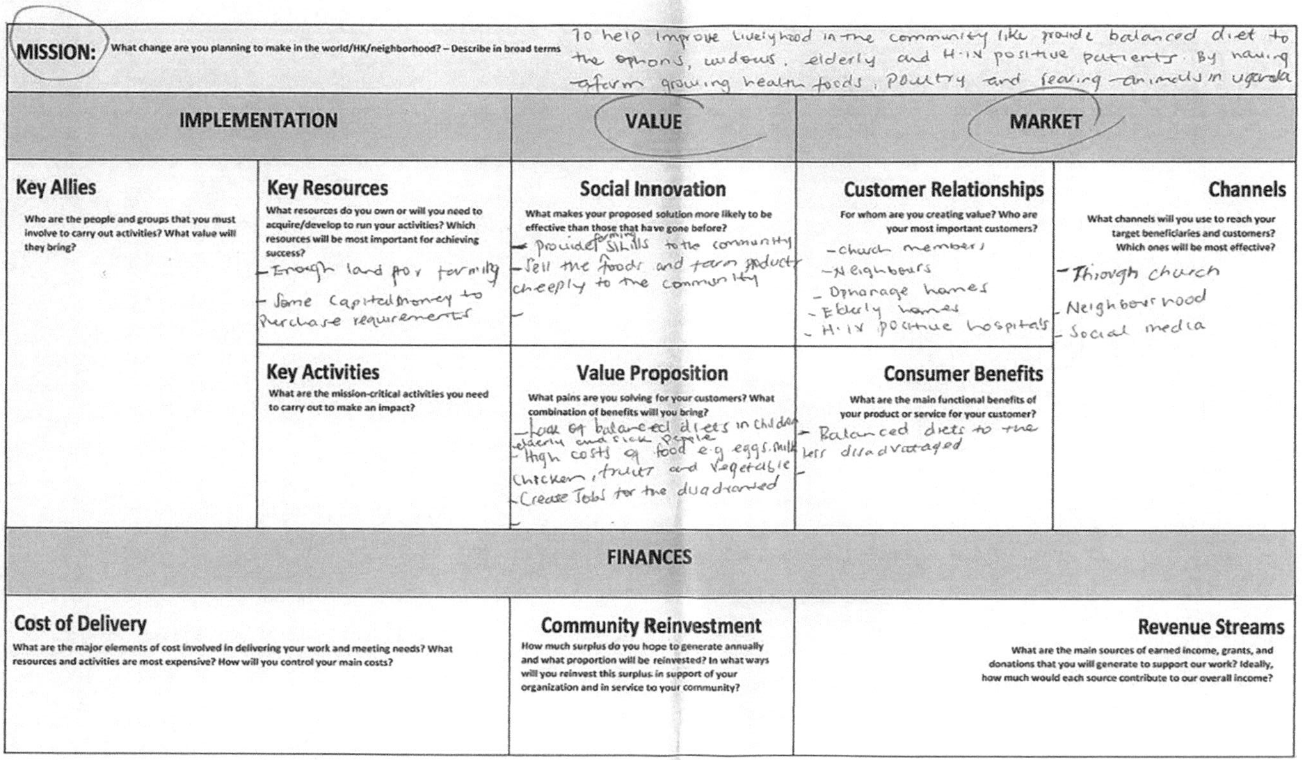

MISSION: What change are you planning to make in the world/HK/neighborhood? – Describe in broad terms

To help improve livelihood in the community like provide balanced diet to the ophons, widows, elderly and H.I.V positive patients. By having a farm growing health foods, poultry and rearing animals in Uganda

IMPLEMENTATION | **VALUE** | **MARKET**

Key Allies
Who are the people and groups that you must involve to carry out activities? What value will they bring?

Key Resources
What resources do you own or will you need to acquire/develop to run your activities? Which resources will be most important for achieving success?
- Enough land for farming
- Some capital/money to purchase requirements

Key Activities
What are the mission-critical activities you need to carry out to make an impact?

Social Innovation
What makes your proposed solution more likely to be effective than those that have gone before?
- Provide farming skills to the community
- Sell the foods and farm products cheaply to the community

Value Proposition
What pains are you solving for your customers? What combination of benefits will you bring?
- Lack of balanced diets in children, elderly and sick people
- High costs of food e.g. eggs, milk, chicken, fruits and vegetables
- Create Jobs for the disadvantaged

Customer Relationships
For whom are you creating value? Who are your most important customers?
- Church members
- Neighbours
- Ophanage homes
- Elderly homes
- H.I.V positive hospitals

Consumer Benefits
What are the main functional benefits of your product or service for your customer?
- Balanced diets to the less disadvantaged

Channels
What channels will you use to reach your target beneficiaries and customers? Which ones will be most effective?
- Through church
- Neighbourhood
- Social media

FINANCES

Cost of Delivery
What are the major elements of cost involved in delivering your work and meeting needs? What resources and activities are most expensive? How will you control your main costs?

Community Reinvestment
How much surplus do you hope to generate annually and what proportion will be reinvested? In what ways will you reinvest this surplus in support of your organization and in service to your community?

Revenue Streams
What are the main sources of earned income, grants, and donations that you will generate to support our work? Ideally, how much would each source contribute to our overall income?

Team 2—Social business canvas.

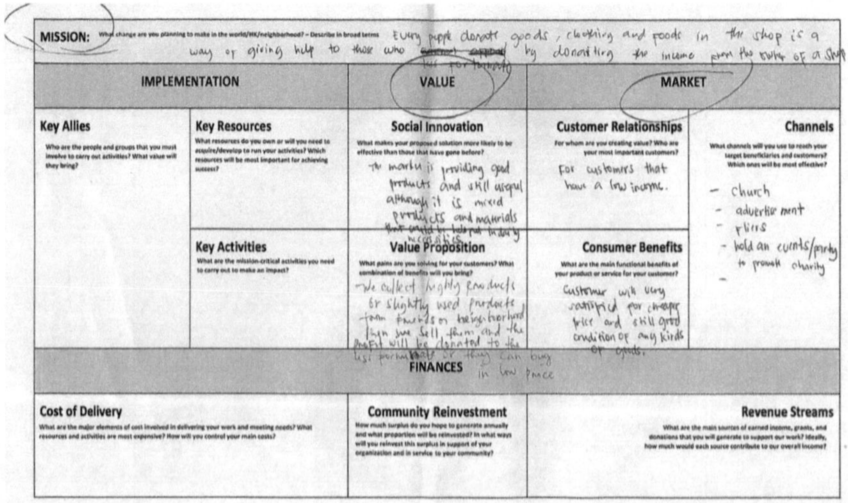

Team 3—Social business canvas.

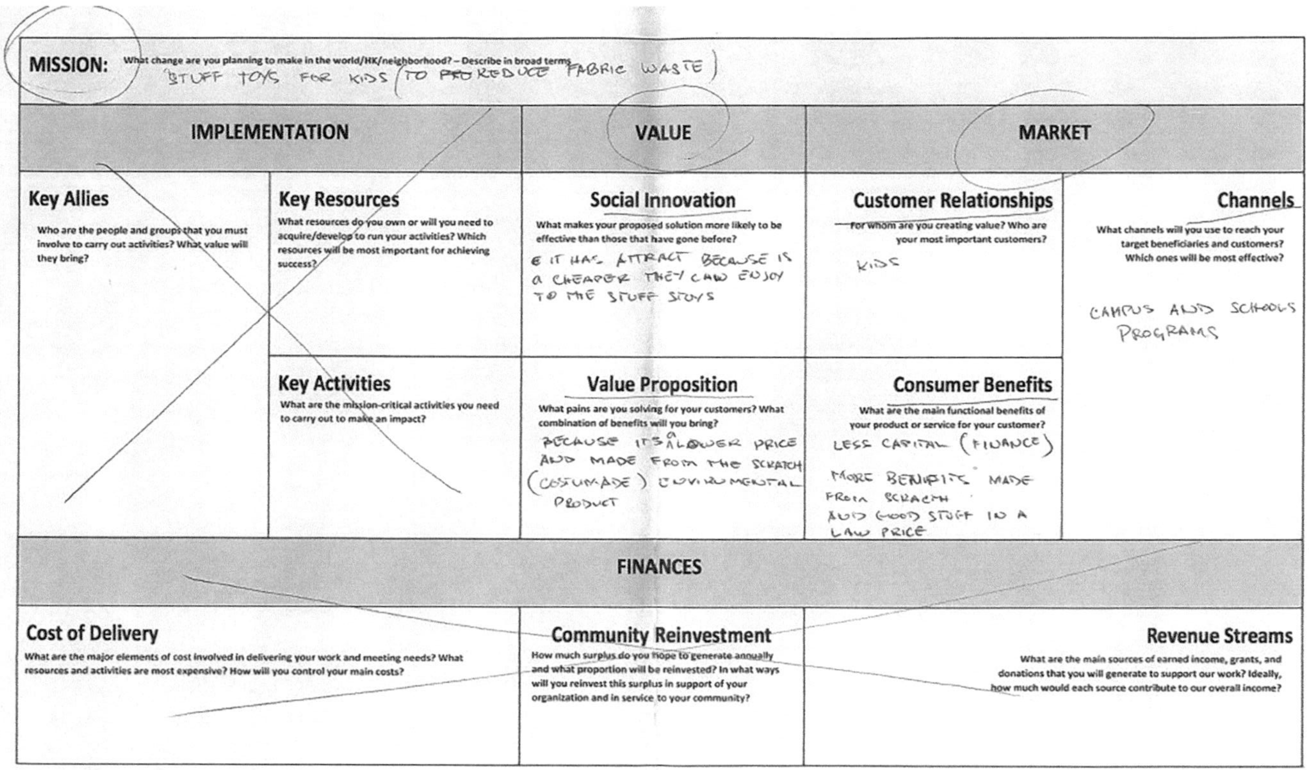

Name Index

A. T. Saral, M. Pavlova, *Education for Refugees and Sustainable Green Urban Development*, Education in the Asia-Pacific Region: Issues, Concerns and Prospects 78, https://doi.org/10.1007/978-981-95-1680-3

Index

A. T. Saral, M. Pavlova, *Education for Refugees and Sustainable Green Urban Development*, Education in the Asia-Pacific Region: Issues, Concerns and Prospects 78, https://doi.org/10.1007/978-981-95-1680-3

GPSR Compliance

The European Union's (EU) General Product Safety Regulation (GPSR) is a set of rules that requires consumer products to be safe and our obligations to ensure this.

If you have any concerns about our products, you can contact us on ProductSafety@springernature.com

In case Publisher is established outside the EU, the EU authorized representative is:

Springer Nature Customer Service Center GmbH
Europaplatz 3
69115 Heidelberg, Germany

Batch number: 09460586

Printed by Printforce, the Netherlands